AWAKEN
YOUR IMMORTAL
INTELLIGENT HEART
A Blue Print for Living in the Now

CARMEL GLENANE B.A. DIP. ED.

AWAKEN
YOUR IMMORTAL
INTELLIGENT HEART

A Blue Print for Living in the Now

CARMEL GLENANE B.A. DIP. ED.

Big Country Publishing, LLC

Awaken Your Immortal Intelligent Heart, *A Blue Print for Living in the Now*
©2016 by Carmel Glenane

ISBN: 978-1-938487-23-1 (print)
ISBN: 978-1-938487-24-8 (eBook)

Library of Congress Control Number: 2015957567
Portions taken from the original publication "Awakening The Intelligent Heart"
©2014 by Carmel Glenane & Lisa Malcolm.
ISBN: 978-0-9873439-3-2 and ISBN: 978-0-9873439-3-2
ID35045058© Styleuneed Dreamstime.com

The author and publisher of the book do not make any claim or guarantee for any physical, mental, emotional, spiritual, or financial result. All products, services, and information provided by the author are for general education and entertainment purposes only. The information provided herein is in no way a substitute for medical or other professional advice. In the event you use any of the information contained in this book for yourself, the authors and publisher assume no responsibility for your actions. Big Country Publishing, LLC accepts no responsibility or liability for any content, bibliographic references, artwork, or works cited contained in this book.

Published by
Big Country Publishing, LLC
7691 Shaffer Parkway, Suite C
Littleton, CO 80127
USA
www.bigcountrypublishing.com

Printed in the United States of America, United Kingdom, and Australia

Books may be ordered through booksellers or by contacting:
Atlantis Rising Healing Centre
P.O. Box 376,
Coolangatta,
QLD, 4225
Australia

www.carmelglenane.com
www.atlantis-rising.com.au
www.senjukannonreiki.com
+61 (0) 7 55 367 399

Because of the dynamic nature of the Internet, any web addresses or links contained in this book may have changed since publication and may no longer be valid.

Contents

ᕲ Introduction ᕲ
Awaken Your Immortal Intelligent Heart

About This Book

This book, "Awaken Your Immortal Intelligent Heart" was written under the "All Love" Guides instructions. This is their book, and now, yours!

The scientific methodology in the first section, "Awaken Your Immortal Intelligent Heart," was contributed by Lisa Malcolm, a research scientist and student of the Atlantis Rising Mystery School. Lisa was being attuned to Senju Kannon Reiki™ in 2011 when I asked her, as a student, what she, as a scientist, knew about the endocrine function of the heart. She began her research based on the transmissions I sent her, and her own study.

The body of writing from Lisa not only includes her impeccable scientific research, but a compilation of the "All Love" Guides channeled writing to her, from me, as well as her own writing.

Part I: Awaken Your Immortal Intelligent Heart

The five activation exercises in the second half of this book are initiations and are to be completed sequentially; set time aside to complete them one by one. Space and commitment is needed, but above all "self love" and a sincere belief that "you" deserve to be loved.

Leave plenty of time to process your experiences. Initiations are just that; mini surrenders. Something must be surrendered before you allow yourself to receive. Counseling and support may be required. Seek professional help if emotions surface that overwhelm you.

As you complete the exercises, the following guide will help you connect the chapters and activations to the Atlantis Rising Mystery School ascension program.

Level 1 - Awakening the Intelligent Heart
The Self Nurturing Energy Belt – Base Chakra

Level 2 - Activating the Intelligent Heart's Endocrine Function
The Self Respect Energy Belt – Belly and Solar Plexus Chakras

Level 3 - Secrets of the Feminine DNA
The Self Energy Belt – The Heart and Thymus Chakra

Level 4 - Activating the Heart's Senses
The Self Realization Energy Belt – The Throat and Third Eye Chakra

Level 5 - Surrender to the Black Heart
The Bliss Energy Belt – Crown Chakra and beyond

Making a commitment to activating your intelligent heart now creates a space of trust for you from "you" to grow your heart's intelligence. For all initiations, grief or sadness can be the flip side. Support yourself in this process; set up a group where you can workshop the initiations. The attendees in my workshops have gained so much by witnessing their own journey being supported by loving group members. Above all keep a diary. Call it "My Heart Speaks" diary, being in the company of your "Higher Self" and any form of protection, white light you feel appropriate, is also important. Your greatest therapist is your own heart. Listen to it, and act on "your" instructions. Feel free to share your experiences.

Part II: The Immortal Woman

Part II of this book is a channeled discourse from The Immortals. It can be read one page a day, meditated on, and brought into being through reprogramming the cellular memory. It was channeled this way, one page every few days, so my humanness could absorb it like a sponge.

All life is from the principle of the "She"/"Her." She is an expression of that principle and is not meant to be a political or social statement about women in our society.

I thank The Goddess of All Light and all beings who have contributed to this channeled discourse.

I give my gratitude and love to all the human souls who have assisted as well. I know this book has changed my life and will yours when you allow its spirit to infuse your cellular memory.

Since the Immortal Woman teachings were first presented to the world, there has been a profound shift in consciousness in the planet. This book has a universal message and has visibly changed all who have read it and absorbed its message.

Every single person who has read a passage or paragraph has been instantly affected. I have witnessed miracles occur in people's lives who described to me that in having the book in their presence, whether it be in a handbag or on a bedside table, has helped them to change their lives. By touching the cover and attuning to the "vibration" of The Immortals your hands can become warm as the energy moves through your body. Little did I know when I first published *The Immortal Woman* how and in what magnitude it would affect all who read it.

All Love, Carmel

PART I:
AWAKEN YOUR IMMORTAL INTELLIGENT HEART

Chapter 1

YOUR INTELLIGENT HEART

When you live through the intelligence of your heart, you are no longer searching for truth; you bring your truth with you into every moment. To upgrade into your intelligent heart is one of the greatest gifts you could ever receive.

An intelligent heart brings others around it into greater coherence and creates a more peaceful, harmonious, and joyous space around you.[1] An intelligent heart increases your magnetism and attracts greater love into your life. Things that once caused you great pain and distress no longer bother you and you no longer have to scramble to attempt to fulfill your basic needs for happiness, security, and a sense of well being.

Mostly, individual lives are out of balance, directed by the mind without the balance of a fully functioning heart. People struggle and experience the frustration of trying to satiate the natural need for love, peace, and a sense of well being that the awakened heart brings, trying instead to fulfill those needs with things like bigger incomes, more gadgets, romance, religion, adrenaline fixes, and "plastic" spirituality. The need we seek to fulfill is satiated truly by the acceptance of our own heart, and by feeling your heart connection to other hearts and realizing your connection to the one heart.

Beautiful things happen when people are in their hearts. The evolutionary shift into our awakened, intelligent, and loving heart is a necessary step in humanity's evolution and it is the solution to many of the global and individual challenges that face us. Our Earth is wondrous and offers all we need for sustainable life. Yet, in the disconnection from our hearts, from each other, from the natural world, and from our own divinity, we have damaged much, perpetuating environmental degradation, social injustices, and feelings of fear and separation. This is simply not possible when your heart is fully functioning, awakened, and intelligent.

Holistically combining ancient wisdom, channeled knowledge with paradigm changing scientific breakthroughs, allows the seemingly impossible reconciliation of science and spirituality to occur as a natural consequence.

All living systems are bound together in energetic interconnectedness, and as each individual awakens their intelligent heart and lives by its wisdom, the coherence of the human energy field increases, that is, the vibration of the collective energy field of humanity's consciousness shifts to a more loving vibration.[2] As each individual moves into their heart consciousness, it becomes easier and easier for others to awaken their own intelligent hearts. Each awakened heart is a catalyst for the awakening of others' hearts.

Our hearts radiate a measurable electromagnetic field that can have an effect on other beings.[3] You may have noticed this yourself, experiencing feeling "good" after being in the presence of a loving, highly vibrational person. Even if you are completely unconscious of this field and your heart's capabilities, your heart can affect the resonance of others around you. Your heart's resonance, as part of the collective resonant field of humanity also effects the resonance of the planet and influences the manifestation of events and circumstances.[4]

The field of human consciousness affects the Earth so much that fluctuations in the Earth's magnetic field at times of mass human emotion can even be detected by National Ocean and Atmospheric Administration's (NOAA) satellites.[5] That the Earth's energetic systems are influenced by the collective consciousness and human emotions has also been confirmed by university studies that report very strong evidence of the interaction between the collective consciousness of humanity and its influence on global events.[6] Other studies in consciousness have conclusively shown time and time again that when a group of people come together with hearts resonating to the higher frequencies of love, events can be influenced.[7]

Each of us individually is responsible for our contribution to the resonance of the collective field and therefore also for the manifestation of events and circumstances. These circumstances can in turn have an effect on us. It is a truly interactive system. You are a unique part of an interconnected whole.

As an individual you experience your best quality of life when your heart is an awakened, intelligent heart, as your heart then leads you to making decisions that support your true happiness. As you live more and more by your heart's intelligence your responses to events and your reactions to emotional impulses become more harmonious and balanced, it flows on to your immediate environment, and your family, your workplace, and your community also become more coherent.[8] Just as your local environment benefits from your heart's awakening, so too the global resonance, as it is enhanced by your love. With the shift of more and more people into heart consciousness, the transformation of the global community from one where warfare, poverty, injustice, and ecological destruction abound to a global community of peace and abundance, where we treat ourselves, each other, the Earth, and all beings with love and respect, it is a natural progression. It is truly our responsibility to ourselves and to the collective, to bring our hearts to the higher resonances of love, power and wisdom, and awaken the intelligent heart.

So just what is an intelligent heart and how do we awaken it?

An intelligent heart has its five senses activated, it balances the endocrine system of your body beautifully, it has its own IQ, the revitalizing, health-restoring Mitochondrial DNA (MtDNA) is activated, and ultimately the intelligent heart is the doorway to the "unified field" of creation.

The heart is the seat of our greatest intelligence and the term "intelligent heart" is not just an analogy, the heart is an actual intelligent organ. The heart has a brain. In fact, more than half of the heart is composed of the same type of neurons that are found in the cerebral system, the "cranial brain" in our heads. The sometimes called "little" brain in our heart contains a circuitry of around 40,000 neurons, as well as neurotransmitters and other cells like those found in the cranial brain. The discovery of the heart's brain has given rise to an amazing new field of study called Neuro-cardiology.[9]

Your heart's little brain talks to you all the time. To begin a conversation with your heart is an intimate discovery of your "self" in its true essence. Imagine a dialogue with an intelligent organ, that only knows one thing, and that is the preservation of your own true self.

It is now well known that in the dialogue between our hearts and our minds, the heart actually sends more signals to the brain than it receives from it.[10] The heart's brain operates independently from the head (cranial) brain. This is clearly evidenced in heart transplants where medical science has never found a way to reconnect the vagus nerve, the nerve that connects the heart to the cranial brain, yet when the heart is in the new body it starts beating anyway.[11] The heart is also the first organ to be generated in the newly developing fetus. The "brain matter" present within the heart; the neurons, neurotransmitters, etc. allow the heart's brain the ability to process information, to make decisions, and just as in the cranial brain, the heart's brain demonstrates a type of learning and memory.[12]

Your heart also has the ability to independently sense. The sensory neurites within the heart's brain allowing the heart to actually see, feel, hear, smell, and taste, when you awaken your heart's senses.

The ancients in their wisdom knew the heart to be the seat of the self, of emotion, thought, will, intention, and the spiritual seat of intelligence and memory.[13] Modern science now, knows that the heart is the main control center of our body. It links our body, mind, emotions, and soul. It sends signals that influence the regulation of all of our body's systems. The heart is also the primary generator of rhythm in our body, and the pulsing rhythms of the heartbeat influence the processes that control your nervous system, cognitive function, and emotion.[14] Both the thoughts of our mind and the health of our body as well as our emotional state can be positively transformed by listening to our heart's intelligence. Our hearts have true intelligence, and by activating and listening to them, we gain access to wisdom, compassion, kindness, and emotional balance in our lives.

One very simple way to begin to awaken to your heart's intelligence is to simply place your hands on your heart. Just put your hands over your heart and let your awareness focus there. With your hands on your heart, feel the love and affection you have for it, this amazing organ that supports your life. Smile to your heart. A smile heals, and smiling to your heart helps develop your ability to give and receive love. Smiling to your heart is like bathing in love. Say to your heart, "*Heart, I love you.*" You may feel tightness in your chest, an expansion of your heart, or a feeling of calm or bliss. You may not feel anything at all. Continue to relax deeply and keep your awareness on your heart, gently letting your heart know, "*Heart, I love you.*" Do this daily, as many times as you feel, because you cannot overdose on love. Smiling to your heart and intoning, "*Heart, I love you*" is a very powerful practice; it allows your heart a new way of functioning.

You are feeling now the precious magic of this gift awakening you and allowing you to really begin to feel the love for an organ you probably didn't really "own." The peace of allowing your own heart to tell you what it wants for your day brings to you a rare sense of ownership of your humanness. This dialogue can begin a new awakening of your true humanness. Your heart's "ears" must be finely tuned for this awakening now, as you bring this gift to yourself. You may like to begin a "Heart Speaks" diary and record in it each day what your heart has to say as you begin to have a conscious dialogue with your heart.

To begin your heart speak conversations is to acknowledge that your intelligent heart has a brain and consciousness which goes beyond anything you could have previously imagined. You are beginning to pour love and light into this part of yourself now, as you awaken to the part of yourself that lies dormant and has been a sleeping giant. Feeling this love and energy brings you a sense of magic and worth, in a way you couldn't imagine.

Allowing this special magic allows you to really bring to your life the knowing that your heart will talk to you in dialogue when you begin to say daily, "*Heart, I love you.*" Saying to your heart, "*Heart, I love you,*" allows you to tap into an undiscovered part of yourself, one, which only knows peace and light. Establishing a dialogue with your heart teaches you to be silent, to be a witness to your heart's birth in your human storybook. "*Heart, I love you*" taps into a rich river of power and light for your self-discovery. Many people feel separate from their heart and find it difficult to begin this dialogue to attune to their heart.

A simple way to begin to reconnect with your heart's intelligence, is to just place your hands over your heart space. Just placing your hands there will bring your awareness to your heart. You don't have to do anything; don't have to force anything, just simply place your hands over your heart. You can do this many times throughout the day. Simply bringing your awareness to your heart will begin to awaken you to its intelligence.

By intoning, "*Heart, I love you*" you are shifting old stuck patterns, opening the pathways for this remembering. The process of allowing this to take place brings you now into resonance with all there is. What is "all there is for the heart?"

All there is for the heart is a pattern of remembering that must take place before any real progress can develop in the heart's awakening. This awakening is a knowing that all of life and you are one, and every part of all life is every part of you. Your awakening is dependent upon this immortal aspect of your remembering. Allowing this remembering brings you into resonance with all there is. All of life and you are one.

You are the sum total of this outpouring of love to you, and you in turn reflect this back, strengthening your heart's capacity to feel even more love. Just feeling this now brings you home safely to all of life.

The joy of allowing the heart to bring you its magic is in allowing the truth of your heart to be revealed to yourself. "Why do you want an intelligent heart?" "What is in it for you to want to embrace this truth?"

Your life force magnifies enormously when this energy is around you, and you must feel the magic, the essence, and the life force of all around you bringing you all you need for your life. Just feeling this essence allows the sacred forces to create with you. The sacred forces correspond to your heart's intelligence, as they are attuning to your heart all the time.

All attunements to the heart take place in the company of mighty intelligences, which are sent to support the soul in her evolvement. This evolvement shapes the destiny of all the forces in the universe as well, and your heart's intelligence grows with all the remembering you are capable of. Believing is the magical capacity allowing the forces of love and nature to combine to bring in magical intent. Bringing in magical intent reinforces the soul's journey to feel and to remember all there ever was and all there ever will be.

It is time now to allow this special magic to bring you all you need for your life of sweet remembering. The past, the present, and the future merge into one, one knowing, one love, and one heart. For, we are all part of the one heart anyway.

Allowing the special magic of truth to emerge brings the intelligent heart's vibration alive. You are now allowing the energy of this mergence to bring you all you need for your life and your world. This moment in your energetic history allows you to have the peace and light you need to create the magical world of "All Love." The light and energy of your magical world brings with it a sense of renewal of every fiber of your being. Sharing this essence and light with all around you speaks now to you of the relationship of the world of truth.

This world of truth allows you now to create with this newly created intelligent heart. For now you are crafting your intelligent heart and bringing it into resonance with all of life. You are embracing the energy, world, and essence of this magic in everything that you do. Crafting this new identity brings you into a relationship with all there is. Your newly crafted intelligent heart allows you now to really release to a part of yourself that only knows one thing, the relationship with yourself. This is a new relationship and its energy must be fostered in your energy matrix right now. Seeing this relationship emerge now allows the light and energy of your renewal to bring you all you need for your love- filled life. Your energy now needs to be created with the power of "All Love."

By being still and attentive to your heart's remembering you are activating dormant cells in your heart. Every cell becomes alive to its potential as the essence of this remembering brings you now into complete harmony with your own truth. For you are a witness to yourself in this remembering and you are bringing to your world your power to love. Your power to love is shaped by the conditions you feed it. What fodder are you feeding your heart right now?

The conditions your mind places on your remembering is an enormous struggle for the heart. You are feeling this power and light begin to awaken you from your slumber and your mind will create every diversion it can think of so it must be put firmly into the background, for then it has no control. You must be very gentle with this part of yourself for the mind needs to be trained and molded. Molding and shaping this part of your consciousness brings you home, to your own special space of purity for all you are capable of.

Right now the expression of what your heart wants must be first acknowledged and surrendered to; you must want something the heart can bring you first, peace, space, order, love, abundance, wisdom, etc. Develop a relationship with your heart, and then you will begin to lovingly allow your heart to bring you all you need for its truth. This essence is shaping your relationship with all of life now.

Our heart's intelligence is the barometer for our emotional, mental, and physical well being. It is important to listen to the heart and monitor its intelligence. Your heart's IQ is measured by the love you are capable of receiving and giving; with the immeasurable love our heart is capable of, why does it stop feeling? It stops feeling when we wall up emotional hurt. By walling up emotional hurt we prevent our heart from being able to see, feel, and remember. What shape is your heart in right now? Ask yourself, "How is my heart's IQ?" How much love am I receiving? How much love am I giving out? By going into a meditative state and going into the heart's intelligence you will find a space within your heart that knows all answers about your heart's IQ.

Now, place your arms out. Feel which arm is heavier to you. Your left arm is your receiving arm and your right arm is your giving arm. Your left arm could feel heavy.

What is it saying about your heart? You may be feeling you are overloaded with the responsibilities of your life that you are now receiving. If your right arm feels heavy, your masculine is overloaded. You are giving out at the expense of receiving. You just keep giving. It's like letting your bank balance go down. One day there will not be enough in the account because you haven't allowed yourself to receive. The energy of truth emerges through the oceans, and we are gathering to remember that we all once came from the oceans. Life comes from water...

Just now feel your body and senses being enveloped by the nurturing releasing ocean. Be bathed in the essence of the oceans, the mysterious depths and life forms which have an intelligence we can only dream of. The essence of the ocean is going to speak to you now so you can remember who you are. You are part of a remembrance process when you surrender to the ocean and marine life. All life surrounds you now and you dip gently into the mother, the sea. Feel now you are under the waves, going deep into a place within yourself where the mysteries of the sea will be revealed to you. The mysteries of the sea bring to you now a special knowing and a dolphin will appear. You will undergo a ritual of release and surrender to the wisdom and remembering of the element of water. You will bring to yourself this remembering. You may ask for your DNA to be recoded to receive your full component of the ascension and mergence into light right now. You are part of the remembering process NOW, you are love.

Chapter 2

ACTIVATING YOUR HEART'S ENDOCRINE FUNCTION

The heart is the master gland of the endocrine system. The endocrine system regulates the body's functions and allows for the assimilation of light. By consciously activating the heart's endocrine function the whole body is strengthened and rejuvenated.

The pituitary gland, located in the head and associated with the third eye was long considered by medical science to be the master gland of the endocrine system. The heart's crucial endocrine function was being completely overlooked. The heart's role in the endocrine system was unknown, even to cardiologists, until the discovery of cardiac hormones made it necessary to "completely revise the concept of heart function." This discovery, published in the *American Journal of Heart Circulatory Physiology (2006)* expanded the field of cardiology and the heart was finally being recognized by medical science to be an imminent part of the "integrated systems of the body including nervous, endocrine, and immune systems."[1]

In fact, the heart's endocrine function is the master of the whole endocrine function! This is the single most important thing to embrace. The heart being an endocrine gland shifts the whole focus away from the mind. Suddenly, now it isn't the mind running the show of human consciousness, it is the heart. By seeing the heart as a master endocrine, you are shifting the focus to the seat of the self, the seat of the soul. When the heart as the master gland of the endocrine system is activated, the endocrine function in the whole body becomes strengthened; this activation has facilitated profound healing experiences.

It is our endocrine systems that regulate most of our body's functions. The glands of the endocrine system communicate via the release of hormones; these are the body's chemical messengers. They transfer information throughout the body and influence almost every cell. The heart is a flexible organ, a living, breathing intelligence, a huge hormonal pumping station. It responds to emotion, sadness, and losses in the heart by pouring more energy into the troubled wound. If the heart goes into shock with grief and loss, the energetic damage can be enormous, as it puts a great strain not only on the heart, but the entire glandular system. The major glands of the endocrine system are: the heart, hypothalamus and pituitary glands, thyroid and parathyroid, adrenals, pineal gland, the reproductive organs, and the pancreas. This system and related hormones have a large influence on us. They regulate our moods, cell growth, development, tissue function, metabolism, and sexual and reproductive functions. So, when you consciously activate the heart as an endocrine gland, the whole body becomes strengthened, and you will feel the energy of it bringing to you your ability to rejuvenate.

Knowing the heart as a pivotal part of the endocrine system is an essential step in allowing your heart's expanded intelligence to bring you into balance. The heart's endocrine system can be opened and awakened through the power of the sun and the moon.

Portals are openings in space and time that connect distant realms. Once relegated to the realms of science fiction, portals have now been recognized by modern science. NASA (National

Aeronautics and Space Administration) have recently discovered hidden portals in Earth's magnetic field creating an uninterrupted path leading from our own planet to the sun's atmosphere, 93 million miles away.[2] Remembered in both indigenous and esoteric traditions, creating portals to the moon and sun in temples and sacred spaces was part of the culture of the ancient world. Priests and priestesses would commune with the moon to bring down power, energy, and light to the human body, restoring the inner equilibrium and allowing the sense of true heart awakening to take place.

The moon acts as a conduit for the heart's endocrine function. The moon's power to create change in the human body is well documented and its great power in relation to our endocrine system is obvious. We have only to look at the reproductive cycles of women to see the strong connection between the moon, whose rhythm the reproductive cycle follows and the endocrine system, which regulates them. The Earth's crust itself bulges toward the moon by around 30 cm as the moon passes over it. And, we have only to look at the cycle of the ocean tides to see another manifestation of the moon's powerful influence on Earth.

The moon's energy brings magnetic energy to the heart as its conductivity allows for a resonance to be set up in the molecular structure of the human body, because the molecular structure responds to vibration through sound frequencies. These sound frequencies emit a charge and the human heart awakens. Allowing the full moon to resonate with your energy field and allowing the heart's endocrine gland to be activated brings you into a state of awareness, for all there is. Right now, activate the heart's great reservoir of power, the moon for restoration and renewal, youth, and inner harmony.

You make a decision to activate your heart's endocrine function by allowing yourself to reflect on what you previously thought your heart was capable of.

Ask yourself now: *"Up until this time, what did you think your heart's primary function was in physical, emotional, or spiritual terms?*

You may like to write down your reflections in "Your Heart Speaks" diary.

Then allow your consciousness to expand by acknowledging exactly what your heart's capabilities are. This expansion of consciousness sets up a new pattern of vibration in your heart and you are then able to receive transmissions from the stars, the sun, and the moon. To enable this consciousness to be activated in you implies that you are ready to receive. By acknowledging to yourself that the heart is an endocrine gland when activated and acknowledged as one, you will begin to receive the energy and abundance from the stellar realms, sun, moon, and the Earth herself.

Traditionally, the moon has played an enormously powerful role in humans' healing and regeneration. The moon brings the feminine into resonance with its truth, and it activates the heart by magnetizing the heart's endocrine function.

When the moon is evoked energetically or by moon bathing, you may feel a tightness or expansion in your chest. Ideally this activation of the heart's endocrine function can be acknowledged as a group. Group consciousness implies that the group brings to the world the ideals shared by that group, and the ideals of the group bring power. You may like to invite others to form a heart activation group with you. The heart's endocrine acknowledgment encourages group energy and consciousness. When distance stops one person from resonating to any group, attuning to the heart as an endocrine gland enables that person to feel closer to the group purpose. The activation of your heart's endocrine function can also be done by yourself as an individual endeavor.

To awaken your heart to its full DNA potential, including both your Mitochondrial and double helix DNA, you must begin now to set up a resonance within yourself by visualizing your heart's Mitochondrial and double helix DNA ready to take full activation.

"I now allow my heart's Mitochondrial and double helix DNA to be fully opened to receive the power of the moon, through activation of my cellular memory for a fully awakened, intelligent heart."

Be still and visualize the moon now as an intelligence. Really feel this intelligence. Right now you need to reflect on your heart, and what it wants.

The sun is a portal to the universe, the moon another portal. Slip into the moon's portal now, and allow the reflective nurturing qualities to give you all you need for your life. Breathe now deeply, powerfully, and rhythmically, breathe in the moon's powers to heal and restore. Feel yourself just slipping into the moon's radiance, be still in this energy. Really feel it.

When you feel complete with this activation, allow yourself to connect through the moon's portal to the sun and feel now the difference in your heart. The moon births you to the sun, and the sun now begins to rejuvenate and bring power and light to you.

Feel this now.

The sun is so intrinsically linked with our endocrine systems that there is even a sub system called the "vitamin D endocrine system." It is named for the vitamin that is provided to our bodies by the sun. Vitamin D is an absolutely essential component for good health, it enables us to maintain good bone health, good immunity, a healthy heart, good cognitive function, and pain control.[3] Healthy functioning of the endocrine system is essential to properly assimilate the Vitamin D that the sun provides and to assimilate the energy and information encoded within the sun's light.

The light of the sun is the basis of life on our planet Earth; it gives energy and information for the biological life forms. Light is essentially made up of energy and information. It is emitted in tiny "packets" called photons. Photons are capable of carrying a great deal of embedded information.

The capacity of light to carry information is commonly utilized in our modern society. For example, optic fibers are tiny hair like glass fibers that are used to transmit light along their length. They transmit information for cabled communications such as cable TV, Internet, and telephones. To transmit data in this way there must be a transmitter to produce and encode the light signal. The light signal travels across a distance, sometimes utilizing signal "boosters" over long distances. The signal is then received and decoded by a receiver; in this case, the telephone, television, or computer acts as the receiver. If the stars and our sun and moon are considered the transmitters and boosters of the system, the biological life forms on Earth can be considered the "receivers" and for we humans, our endocrine system with our hearts as its master are the decoders.

Our sun transmits an enormous amount of energy and information and it is we, the biological life forms on planet Earth, who are the receivers. It is our endocrine systems with our hearts as its master that are most strongly associated with receiving and decoding the information embedded in our sun's light.

Humanity's own technical capacity to embed information in light is increasing rapidly. Scientists have recently discovered a way of twisting light into a DNA-style helix shape. These helical-shaped beams of light are capable of transferring information at an incredible 2.5 terabytes per second; that is around 85,000 times faster than the average broadband cable. It is quite phenomenal when you think of the amount and scope of information that could be transmitted in this way. A scientist associated with the project said, "That's the beauty of light; it's a bunch of photons that can be manipulated in many different ways at very high speed."[4] There is also a new technology where wireless Internet data is transmitted by being embedded in the light from a normal LED light bulb.[5] Even the light from your average desk lamp has the capacity to transmit large amounts of data!

Imagine then how much more information is transmitted by a far greater and more powerful transmitter, our sun; the basis of life on the planet. Imagine again the astronomically enormous amount of information that could reach us from the stars. The Andromeda

Galaxy is 2.5 million light years away; it is the farthest light visible to the naked eye. In fact, it is possible to receive information and energy from the entire universe. This knowledge was confirmed by science when in 2011 astronomers at Australian National University's Research School of Astronomy and Astrophysics discovered proof of a vast filament of material that connects our Milky Way galaxy to nearby clusters of galaxies. These galaxies are similarly interconnected to the rest of the universe. A scientist on the research team said of the discovery, "What we have discovered is evidence for the cosmic thread that connects us to the vast expanse of the universe."[6]

Don't underestimate the stars' energy, as they are living beings, with life cycles lasting billion of years. The stars' energy coming to Earth is filtered by the aura of the sun. It is the consciousness of the civilizations themselves that embed information in the light of the sun and carry it to your spiritual, etheric body, and to your physical body via your endocrine system and DNA.

When you acknowledge your heart as an endocrine gland and allow its full endocrine function, you are doing something that is essential for a fully functioning body, mind, and spirit. The endocrine function of the heart cannot be overstated in any healing work. Breathe in your capacity to feel that statement.

When our endocrine systems are functioning fully we can receive and decode the full spectrum of light properly and the information embedded in it can therefore be utilized. We already know that a well functioning endocrine system is an essential component of good health for the individual. We must also realize that the fully functioning endocrine system benefits humanity's collective consciousness as well. It has been suggested that the malfunctioning of the human endocrine system causes humanity to be unable to fully assimilate the information embedded in the sun's light and contributes to the fear-based paradigm existing in human society on Earth today.[7] The range and integrity of the code we are able to assimilate, forms part our collective experience and as more individuals activate their endocrine systems, we will receive the full range of light into the unified field of collective consciousness, forming critical mass and transcending into a more loving world.

Chapter 3

SECRETS OF THE FEMININE DNA

Our cells contain Mitochondrial DNA; it comes only from our mothers. It is our direct feminine lineage and it is the power source of our cells. By consciously activating your Mitochondrial DNA (MtDNA) you tap into a wondrously rich reservoir of raw power and your body becomes rejuvenated and charged with source energy.

The Mitochondrial DNA is the trigger for the complete restructuring of your humanness. This new definition of being human is now being offered to you, as you demand to have "all the information" that being human offers. When your heart's intelligence is fully activated and awakened, this remembering becomes possible.

Whether your gender is male or female, your Mitochondrial DNA (MtDNA) is inherited only from your mother, unlike the linear, double helix DNA that is inherited from both parents, with one strand coming from your mother and one strand from your father that is located in the nucleus of the cell and is thus called nucleic

DNA. The Mitochondrial DNA is circular and located in the cellular cytoplasm and comes only from the female.

The Mitochondrial DNA is the direct feminine lineage of humanity.

The Mitochondrial DNA powers our cells.

Scientists have traced the lineage of our MtDNA all the way back to a woman who lived in Africa an estimated 200,000 years ago. She is known as *"Mitochondrial Eve."*

Mitochondrial Eve is our great grandmother; she is the common ancestor of all of the human beings existing on the Earth today. This does not mean that Mitochondrial Eve was the first woman alive, nor does it mean that she was the only woman alive at that time. What it does mean is that humanity's feminine lineage, the common great, great, great, etc. grandmother of the present day human population was a black African woman, and we've all inherited her Mitochondrial DNA, thus she is known as Mitochondrial Eve.[1]

The Mitochondrial DNA holds the secrets for longevity and a disease-free life. This DNA is now being recognized as having the power of powers to create a world of truth for humanity. Embedded in the MtDNA is the wisdom, power, and love of the source energy of humanity's remembering. When we can attune to our MtDNA, we are connecting to the source energy of humanity's remembrance, bringing wisdom, power, love, self-nurturing, and self-healing.

As well as holding information for the feminine lineage of humanity, the Mitochondria provide the energy for life. The Mitochondria are the power packs of our cells providing the body's energy requirements. Mitochondria transform physical energy in the form of the food we ingest and the oxygen we inhale into a chemical energy form – ATP. ATP is a fuel that can be easily assimilated by the body's cells; by activating your Mitochondrial DNA you tap into an abundant source of raw energy, bringing power and light to you, rejuvenating your cells, and providing powerful source energy.

ATP is a chemical called Adenine Triphosphate; it is the energy currency of life. It is not only human cells that are fueled by ATP, all complex plants and animals are fueled by this energy source of life. That is, apart from single-celled organisms (amoeba), all life on Earth is powered by Adenine Triphosphate.[2] In luminescent plants and animals it is the reaction of ATP with a molecule called luciferin that creates the light allowing them to shine. In the oceans around 80% of animals are luminescent.[3]

The energy source for the cells of complex life forms on Earth is produced by the power pack of our cells, the Mitochondria.

When you consciously communicate with the Mitochondria and consciously activate your MtDNA, you give yourself a direct line into the universal energy source. Activation of the MtDNA brings to the surface an undiscovered source of raw power. The DNA activation is like tapping into a rich reservoir of raw power. This raw power brings so much energy to the whole nervous system. It is very important to keep focusing on just how magical and life changing your world is becoming with the source energy of this DNA.

All cells (except red blood corpuscles) must have at least one Mitochondria to provide it with energy and "life." Some cells such as skin cells, usually have only one single Mitochondria, while some have hundreds and others such as our heart cells have thousands and thousands. The most densely packed mitochondrial area of our bodies is the heart. The mitochondrial material in our hearts is so highly concentrated, it has been estimated to be around 40% of the total matter in our hearts.[4]

Our hearts are jam packed with this raw energy source and just focusing on its source power will give your heart strength. The heart is flooded with energy and power, and the heart as an endocrine gland becomes activated.

The energy of this activation brings to the surface much suppressed grief as you begin to transit into a new view of your reality and its potential. You can begin to feel the power of this, as an atomic energy, exploding the power and light throughout your whole body. Feeling your body becoming charged up with source energy and power, allows you to really feel the magic of "All Love" in your life. Your world becomes one, with the source energy of raw power for your body. Feel this now.

Allowing this feeling brings you closer to the power of the ancients who knew this MtDNA and used it to bring might and power to their world.

Shaman and indigenous healers can take their consciousness to the molecular level and can bring back from these journeys, images and information including technical information previously unknown to them from the MtDNA.[5] As well as being known by some spiritual practitioners, this has been confirmed by anthropologists studying the Shaman of the Amazon, who have also confirmed that when the Shaman speaks about the importance of restoring the feminine life force, they are talking about the Mitochondria.[6]

The circular shape of the Mitochondrial DNA resembles that of bacteria. Bacteria can store mega amounts of information. Humans are now beginning to understand and utilize the data-holding capacity of bacteria. In China, USB-style memory sticks are being developed that use bacterial cells to store mega amounts of data. In these data storage devices, just one gram of data-storing bacteria can hold the same amount of data as around 450 x 2,000 GB hard drives.[7]

It is thought that billions of years ago these little, "a billion would fit in a grain of sand" Mitochondria, were independent bacteria, which adapted to live inside larger cells in a symbiotic relationship. It was this symbiosis that allowed life to evolve beyond single- celled amoebas to the variety of complex life forms we see on Earth today. Life on Earth may never have evolved beyond single-celled amoeba if not for Mitochondria.[8]

With the accomplishments of providing the global energy currency that powers life on Earth, as well as facilitating the evolution of all complex life forms attributed to it, you'd expect the

Mitochondria and its DNA to be recognized and celebrated. Yet this is not case; in fact the Mitochondria have been described by a leading researcher as being a "badly kept secret" by 'the clandestine rulers of the world.[9]

The Mitochondria, however, did reach some sort of celebrity status, featured in the *Star Wars* series. "Midi-chlorians" are acknowledged by *Star Wars* creator George Lucas to be based on Mitochondria. In the movies, it is taught by Master Qui-Gon Jinn about the "Midi-chlorians" that; "Without the Midi-chlorians, life could not exist, and we would have no knowledge of the Force. They continually speak to us, telling us the will of the Force. When you learn to quiet your mind, you'll hear them speaking to you."[10]

Biologically the "Midi-chlorians" were described in the movies as being "intelligent, microscopic life-forms that serve as organelles within all living cells, existing in a symbiotic relationship with the beings they inhabit and comprising a collective consciousness amongst themselves. They allow for the connection with the energy field known as the Force. This creates the communication with the will of the Force.[11] The famous line from the movie series *"may the Force be with you"* enthralled the imagination of a generation and still does today.

With the MtDNA activation you will begin to experience a worldview that encompasses light and power. Energy you did not believe you had will be made available to you. People who have had the MtDNA activated perform heroic deeds in saving others. The surge of power to fight for the lives of others in a humane way creates the activation. There are a number of well-documented cases where during a crisis, people have suddenly found themselves with superhuman strength to facilitate rescue and save lives.

In 2008, a man was trapped under a Chevrolet Blazer. A Florida firefighter suddenly had super strength and he lifted the large vehicle 30 cm off the ground. The other firefighters, were then able to rescue the trapped driver from beneath it.[12] In another such incident reported in *The Scientific American*, a man rescued a cyclist from under a car by lifting a Camaro weighing around 1300 kg and holding it in the air for around 45 seconds. It was a heroic act of super strength, and the man said that he could not repeat it.[13] One man who was trapped under his vehicle after it had fallen on him while changing a tire was saved by his mother. The mother, a woman in her fifties lifted the vehicle off her son, and held it 4 inches off the ground for 5 minutes until neighbors came to her aid and reinserted the jack.[14] Feats of super strength such as these were reported in the *Journal of Physiology* as being "expressions of strength that we are always capable of."[15] You are always capable of great feats, it just sometimes takes a crisis to activate the Mitochondria for love to perform miracles.

Ancient spiritual disciplines such as Qui Gong activate the Mitochondria and these masters also demonstrate feats of incredible strength and power. Many also experience great longevity as a result.

Cells remain young and active with the Mitochondria maintaining the density of their numbers within the cell and therefore can readily meet the cell's energy requirements while the structural integrity of the MtDNA is maintained. As people age or become sedentary, the number of Mitochondria within the cells tends to decrease therefore causing fatigue and low energy. Exercise in general is a way to increase the amount of Mitochondria in the cells. When the Mitochondria recognize that the body needs more energy they will replicate themselves, increasing their numbers within cells that need the extra energy. Recent studies show that exercise may minimize, and even reverse, age-associated declines in mitochondrial function.[16]

The Mitochondria is the place where metabolism happens. It is the increase in the numbers of Mitochondria that occurs with exercise that can lead to weight loss. The number and activity of the Mitochondria within our cells determines our metabolism. When your Mitochondria are in low numbers or aren't working properly, your metabolism operates less efficiently, resulting in the food we consume being stored, rather than metabolized. The Mitochondria convert the food we consume into energy. So, to lose body mass the

Mitochondria must transform the energy stored in physical form to the chemical energy form of ATP that can be consumed by our cells in their activities. The greater the workload of the cells, the more Mitochondria required to supply energy to the cell. Thus, muscle contains more Mitochondria than fat cells, as they are more active. When the body is more active and requires greater amounts of energy the number of Mitochondria increase by replicating themselves and the rate of metabolism is increased.

For the Mitochondria to replicate themselves in a healthy way, it is necessary to have the MtDNA with its integrity intact in order to replicate a healthy Mitochondria. It is mutations within the MtDNA that cause us to age. MtDNA is without the protective coating that the nucleic DNA enjoys and therefore can be more readily damaged by free radicals. It is quite sensitive and damage to the MtDNA is increased by stress, emotional imbalances, nutritional deficiencies, and exposure to chemical toxins, including some of those toxins found in everyday products. As humans age there tends to be a greater amount of damaged MtDNA within the cells. Fortunately, each Mitochondria has more than one copy of the MtDNA. When the Mitochondria replicate, in their innate wisdom, the most complete, intact, and undamaged MtDNA is chosen to be copied. When the mutation of the MtDNA within a cell reaches a certain rate and there is not enough Mitochondria with intact and undamaged MtDNA for replication within the cell, the remaining Mitochondria cannot provide sufficient energy for the continued cell survival. So the cell in a healthy system dies by the voluntary process known as apoptosis. The material within the cell is then recycled by the body. This prevents illness.

However, when this system fails and apoptosis doesn't occur, the damaged cells will continue to replicate, resulting in good health being compromised. (Allowing cells to die and be recycled when they do not have enough Mitochondria with healthy MtDNA to replicate is a part of good health.)

Increasing and keeping a high number of Mitochondria within your cells and consciously repairing and maintaining the integrity of the MtDNA is part of the secret of health and longevity.

Chapter 4

THE SENSES

Your senses activate your intelligent, loving heart. The five primary senses relate to the elements and to the energy belts of the body. Determining the sense your heart resonates with most strongly helps to awaken your intelligent heart.

Right now birthing the heart is the greatest act of courage any human can give themselves, as you are making a statement about your right to live on this planet in such a way that reflects truth. This is a time now to put aside any fears about limitation in your life as you carve out your truth. Your truth is a weapon for you to carve, and your truth brings with it power and light. Power and light bring the sense of magical renewal for all of life. All of life respects you now, as you bring to your world this act of remembrance. Just feeling now this act of remembrance allows you to really focus on the light of your heart growing for all there is. The light of your heart grows for all to see.

Allowing this light to merge with you now feeds even more light and power into your heart. Your heart becomes fueled with this light and power. Bringing the light and power to your heart allows you to really own a part of yourself that has been lost, disconnected, and shut off from remembering. You are allowing the power and essence to bring you all you need to feel, see, taste, touch, and hear your truth. For when you activate your heart's senses you have a fully intelligent, fully awakened heart to love.

The five primary sensory perceptions of sight, hearing, touch, taste, and smell are the doorways through which we experience the outside world. They are our body's way of making "sense" of the world. Our inner thoughts and feelings are very much influenced by what our sensory experience is at any given moment. What we think and how we feel are quite often a response or reaction to what our five senses are telling us. What you are touching, tasting, seeing, hearing, and smelling to a large extent determines what you will be thinking and how you will feel. Psychologists aptly refer to the sensations as the raw materials of mental activities.

The senses have the power to bring us into the present moment and to the intelligence of the heart. Being conscious of our sensory experience brings our awareness into the present moment. Realizing and being conscious of your heart's ability to truly sense, to actually see, taste, touch, hear, and smell, awakens your heart's intelligence.

The heart itself has sensory perceptions; it can sense and feel. Within the nervous system of the heart there are tens of thousands of sensory neurites. These neurites are tiny projections from neurons. Neurons are the nerve cells that are the basic building blocks of the nervous system. They are highly specialized to transmit information throughout the body. There are different types of neurons; there are motor neurons that transmit information to the muscles, interneurons that transfer information between different neurons in the body and sensory neurons that receive information from the sensory receptors. Sensory receptors (neurites) exist in the tens of thousands within the heart. They perceive sensory information coming into the heart from the outside world and from within the body. The heart translates the feelings and sensations into neurological impulses and processes the information internally. Once it has processed the information, the heart communicates this with the cerebral brain and sends the information to it.[1]

By bringing your awareness to your heart, with the knowing that your heart has the sensory neurites to truly perceive sensations, you awaken your heart's intelligence.

Ask yourself:
What am I seeing now? What am I touching now? What do I smell now? What am I hearing now? What do I taste now?

Be aware of the sensations you are experiencing right now. Simply experience and appreciate what your current sensory perceptions are. As well as sensing with your eyes, ears, nose, tongue, and skin, allow your consciousness to become aware also of what your heart is sensing.

The sensory organs, the eyes, the ears, the skin, the nose, and the tongue transmit only a particular array of the available sensory information. Our eyes can only process information from the visible range of light, a rather narrow band within the electromagnetic spectrum. The remainder of the spectrum, the longer wavelength, lower frequency signals of infrared, microwaves, and radio waves are invisible to us. As are the higher frequency, shorter wavelength signals of ultraviolet, x-rays, and gamma rays.

Our ears perceive only a fraction of the range of sound waves. Sound waves are measured in Hertz (Hz), the number of vibrations per second. Human ears can decode sound waves of between about 20 vibrations per second and 20,000 Hz. Tones between 4 and 16 Hz are beyond the range of normal human hearing, but can be perceived by the body's sense of touch. Many animals hear, see, smell, and "sense" things that humans cannot. Dogs have much sharper hearing than humans and can detect a greater range of sound, up to around 60,000 Hz.

Dolphins emit high frequency waves of up to 150,000 vibrations per second for echolocation and also make lower frequency sounds that can travel farther for social communication; these are around 75Hz and are audible to humans.[2]

There are many frequencies that are beyond the range of usual human sensory perceptions. When you acknowledge your heart, you begin to see, hear, touch, taste, and smell with new awareness. Your heart's senses can attune to frequencies and perceive information outside the normal range of the other sensory organs. Your heart's eyes see more than your human eyes see. The heart's intuitive perception connects to a field of information beyond normal conscious awareness. It is the heart that is the first part of the body to react in moments of what is commonly termed extra-sensory perception, and scientifically rigorous research has shown that the heart's electromagnetic field is measured as being the first part of the body to show a reaction to a future event, before the event has actually happened. The heart scans the future and it receives, decodes, and responds to intuitive information and sends it to the cerebral brain, the second part of the body to show a reaction. The electrophysiological evidence shows that the heart is intrinsic to our intuitive process.[3]

Your heart is truly your oracle and the essence of oracular ship is a fully awakened and intelligent heart. This is the gift to yourself right now.

The frequency of this love just grows and grows. Feeling this love grow and grow feeds light into your aura. Your aura becomes bathed in incandescent light, a beacon of power, light, and energy. This power, light, and energy embrace the one heart. The heart of universal knowing, allowing this heart to bring you all you need, shows you that you are capable of feeling the mystery of oneness in everything. This oneness is contributing to your overall well being. A sense of joy-filled abundance becomes yours, and you are now attaching to a part of yourself that only knows love, strength, and power.

All around you are examples of this living oracular ship. The world of perfect love is to live in harmony with the essence of all around you. The essence of love brings the mystery right into your consciousness. When love grows in your heart, your consciousness expands to take in even more light. This light allows for the mystery to unfold and you become a living oracle. Feeling this allows for the special magic and oneness to embrace you. You become one with the mystery and the mystery is yourself. This is your time to really allow this magic to become one with you. You are love.

Of the five primary senses; touch, taste, smell, sound, and sight, each one relates to one of the five elements. The physical world is made up of these elements of earth, air, fire, water and the quintessential fifth element of ether. Ether is the subtlest of the elements and is beyond the material world. Earth is the coarsest of the elements and is made up all the finer four elements. Air, the next coarsest material, is made up of all the finer elements and so on. Each element contains all those that are finer that it. The elements can be considered both as substances and as categories of sensory experience, as it is the elements that act as the medium for the experience of the sensations.

As well as the relationship of the elements to the senses, each element also corresponds to one of the five platonic solids. The platonic solids were named after Plato, the Greek philosopher and mathematician who lived around 400 BC. Plato founded an academy in Athens and is credited with creating the first prototype of a university. Although the solids are named after Plato, he was not the first to identify them. The knowledge of the geometric solids was handed down from the Egyptian Mystery Schools. Plato was a student of the Greek philosopher Socrates and it is Socrates who is credited as being one of the founders of western philosophy; he was also a student of the great philosopher, mathematician, and mystic, the Egyptian-trained Pythagoras.

The five primary platonic solids: the cube, tetrahedron, octahedron, dodecahedron, icosahedrons, and their combinations create all molecular structures. These five geometric forms are the building blocks of life and all forms as we know them, from the tiniest molecule to the stars and galaxies; all are comprised of these specific geometric structures. The *cube* corresponds with the element of *earth*. The *tetrahedron* relates to the element of *fire*. The *octahedron* relates to the element of *air*. The *icosahedron* relates to *water*. The *dodecahedron* corresponds to the quintessential element of *ether*. Pythagoras and modern science suggest that the dodecahedron embodies the structure of the entire universe.[4, 5]

We all experience things in different ways and just as each person has a particular sense that their heart will be most activated by, each person relates in different ways to the different elements. Usually a certain sense is dominant and there is an affinity with a particular element. This was known since ancient times and is explained in one of the oldest texts known to humankind, *The Kore Kosmou*. *The Kore Kosmou* translates to Virgin of the World; it is considered to be one of the earliest ancient hermetic texts included in *The Stobaeus Manuscript* by Johannes Stobaeus in the 5th century AD. Its creation is attributed to Hermes Trismegistus, the Greek name for the Egyptian God Thoth. Four of the elements—earth, air, fire, and water—were named in the *Kore Kosmou*.

In the text Isis is speaking to her son Horus and she says:
Of living things, my son, some are made friends with fire, and some with water, some with air, and some with earth, and some with two or three of these, and some with all. And, on the contrary, again some are made enemies of fire, and some of water, some of earth, and some of air, and some of two of them, and some of three, and some of all…It is because one or another of the elements doth form their bodies' outer envelope. Each soul, accordingly, while it is in its body is weighted and constricted by these four. Wherefore, it is natural that the soul should have affinity with certain elements and aversion for others.

Which particular sense is the primary receptor for the activation of your heart's intelligence?

To discover your heart's primary sensory receptor, attune to your higher self. Attune to your higher self by seeing it as a golden ball above your head. By activating your higher self, your soul or unconscious, you will be guided by it to find the primary receptor: sit back, and take a few deep breaths, to relax and ground you. As you breathe, feel all the tension leaving your body. Your body is now very relaxed:

Imagine someone hands you a beautiful rose. *Which sense first draws you to the rose?*

Do you simply look at its beauty or do you put it to your nose to smell its perfume? Or are you wishing to put it against your cheek and feel the softness of its petals. Is it the gentle music the romantic rose reminds you of? Or do you think of the last time you had rose water, rose petal tea, or gulab jamin (an Indian dessert with rose petal syrup)? Is it the taste of the rose you most react to? *Which sensory perception showed as the one that the awakening of your heart's intelligence is most resonant with?*

Were you drawn mostly to the visual beauty of the rose? Wherever we are, we can find spots of beauty to appreciate. If your heart's intelligence is most activated through the sense of sight, look at the beauty all around you. More than half the body's sense receptors are located in the eyes.[6] Spend time looking at things that you find beautiful, the trees, flowers, the sky at sunset, the water, or whatever it is that gives you that feeling of looking at something truly beautiful. Let the beauty that you see all around you sink into your heart, and let your heart see the beauty.

If it was the scent of the rose you were most drawn to and your heart's intelligence is primarily activated by the sensory perception of smell, then seek out pleasant aromas and create pleasant aromas in your own environment. You can do this with essential oils, room sprays, flowers, spices, herbs, fresh baked bread, or whatever it is that you enjoy smelling. Smells are very good at triggering memories. When memories are triggered by the olfactory sense of smell, they tend to be more emotional than the same memory triggered by another of the senses.[7] The memories of particular aromas are also said to be less influenced by the passage of time then auditory or visual memories.[8] Let the smells your olfactory senses perceive activate your hearts remembering. A person, whose heart was activated by their sense of smell, probably coined the saying "stop and smell the roses."

Was your heart most receptive to listening to some gentle music that the rose attuned you to? Music is a powerful mood enhancer; it can relax or excite the listener depending on its qualities. Some traditions teach that it was sound that created the universe. Sound does have the ability to create form. Cymatics is a field of study that focuses on inert materials being animated by the sound waves of pure tones. When the materials used in cymatics (usually powder, pastes, or liquids) are "treated" with sound waves, they form into flowing patterns. These are the same geometrical patterns that can be seen in nature and in the sacred art and architecture of the world's wisdom traditions.[9] Writer and philosopher Goethe said that "sacred architecture is frozen music."[10] Water molecules when exposed to beautiful music also form into harmonious, structured geometric forms. Piano notes when made visible using cymascope technology show as "beautiful holographic bubbles, with shimmering kaleidoscopic patterns on their surface."[11] Sound frequencies are related to color, form, and states of consciousness.

Listen to sounds you love whether it is beautiful music, the sounds of birds singing, water bubbling over rocks, the breeze fluttering the leaves of the trees, the sounds of the waves on the beach, children playing, or whatever it is that you enjoy listening to. Listen to the sounds that you love and hear them with your heart.

If you were most drawn to placing of the rose against your skin to feel the softness of its petals, then touch is a primary sensory receptor for your heart's remembering. Touch is an elementary sense that all the other senses evolve from. When the eyes see, they are essentially "touching" the light waves to receive the information about them. Similarly when the ears hear, they are "touching" the sound waves to determine their characteristics. Your sense of touch is distributed throughout the body and nerve endings in the skin and in other parts of the body transmit the sensations. Hair also magnifies the skin's sensitivity. Some parts of the body have more nerve endings than others and so are more sensitive to touch. The fingertips are some of the most sensitive areas of the body. Hugging loved ones is a powerful way for touch to open the heart. You can also wear materials that feel lovely against your skin. Take a warm bath and be conscious of how the warm water feels around you. Be keenly aware of the feeling of the different textures your sense of touch experiences throughout the day. Get a massage. Patting pets is another great way to activate your heart's sense of touch. When you touch something, let it touch your heart.

If it was the desire to taste the rose that was the most strongly aroused in you, then taste is a primary sensory receptor for your heart's remembering. The taste buds are tiny cells of the tongue; they can perceive the tastes of sweet, sour, salty, and spicy. Make an awareness of tasting and savoring life. Taste the air, taste the water, and taste all of life. Feel it on your tongue. Let your heart experience its taste. When you eat food make an awareness of tasting and savoring it. Don't just eat absent heartedly, gulping your food, eat with your heart's awareness.

The remembering of your heart's intelligence will be activated by bringing awareness to the sensory perception of the heart, focusing on the type of sensory experience that most resonates with your heart will bring it to its remembering. The capacity to really believe in your heart's remembering will develop your receptivity to all around you. You will be allowing your own heart to shine with energy and radiance all the time. The energy and radiance of a fully awakened intelligent heart brings you into a space of remembering all you ever were, as your vibration takes you higher into yourself. Feel your energy right now; you are resonating with a new frequency that brings light and energy to your remembering.

This is a time to really feel the essence, energy, and life force of this remembering, bringing you all you need right now. Your truth becomes revealed at this time and your truth will bring you home to yourself. You are allowing the special light and energy to grow your capacity to feel and to experience this feeling in a new and powerful way. You are feeling now, your remembering connects you deeply to your ability to feel. This feeling brings you into a space of absolute joy in the capacity to know that your heart can see, feel, taste, touch, smell, and have this remembering.

The remembering is just so powerful now and you need to really feel your truth, in its power and light. You are opening up to the world of absolute joy in this remembering process now. Allowing this to happen opens you even further to your truth. Your heart needs to really allow you to experience the magic of this remembering. Bring now your essence to your heart in this remembering. Allowing the forces of nature to assist in your heart's remembering brings you to a space of pure delight in the process of being human. You are feeling this light and power resonate through you as you open up to the world of the most spectacular light show in your consciousness. You are asking now for this energy to be around you and your whole molecular structure, as you bring to your world this remembering. Believing in this precious remembering will allow the space, light, and magic to envelope you, as you create in this remembering your power and your light.

You are feeling this energy surround you now.

Allowing the spirit of complete trust in the surrender process brings your heart into alignment with its true purpose. It shines and begins to feel welcomed in the human journey. This is a sacred time in your earth journey right now, as you bring the peace and remembering to all that you do, think, and feel.

As the newly awakening heart begins to stir from its slumber, it is like a tiny kitten opening its eyes to the world. Its eyes open for a few seconds, close, open a little longer, until the kitten is fully seeing. It is just like that with our heart. The frequency of "All Love" brings to humans now the energy of pure delight in the process of being human.

Chapter 5

CREATING FROM YOUR HEART

Your heart has the biggest energy field of your body and at the center of its center is a point of singularity that is both infinitely small and infinitely dense. It is a doorway within your body to the universal vacuum; a doorway to the infinite.

When your heart unfurls into its fully functioning, awakened, intelligent self, it can come into harmonious resonance with the unified field and you truly can become one with the "field." This field of oneness has gone by many names, the ether, source, the vacuum, the zero point field, or put quite simply "space," it permeates all things. Our physical world that seems so solid to us, our tables, chairs, walls, bodies, everything in our physical manifest world is 99.9999% space. The atoms themselves are 99.9999% space. The interesting thing is that "space" isn't empty, in fact "space" is incredibly dense, it is so dense that if it were at the same density as space, all the actual matter in our entire known universe would fit into an area of space just 1 cm cubed.[1] This incredibly dense space, permeates ALL things.

We are now coming to a conscious understanding of the workings of this field of "oneness with the all." We stand in a timeless time in humanity's collective consciousness, when the hints left to us by those who attained this knowledge, through the deep intuitive wisdom of our hearts which has been expressed by sages and mystics throughout the ages, is now merging with the current level of scientific knowledge. As a result of, the breakthrough discoveries and unifying theories of Nassim Haramein[2, 3] we are for the first time in humanity's modern history now at a place where the wisdom of our sages and our ancient symbolism is in attunement with our mathematical equations and scientific knowledge.

Every atom in the universe is connected to every other atom in the universe by the very structure of space itself. The 0.00001% of the known universe that isn't currently thought of as space are the protons, neutrons, and electrons that make up the atomic structure of atoms. The protons and neutrons are within the nucleus of the atom and electrons are in orbit around the nucleus. If we scaled the size of the densest of atoms and made the nucleus of this atom the size of an apple, the electron orbit would be two football fields in diameter.

The atom itself has a toroidal (doughnut shaped) energy field. The torus is the fundamental form of balanced energy flow found in all continuously sustainable systems of nature, in all scales, from the micro-atomic to the macro-galactic.[4] The torus of the atom, embedded in the torus of the cell, embedded in the toroidal energy field of the seed, the human, the tree, etc. are embedded in the energy field of the Earth, the solar system, the galaxy, the universe...The toroidal energy field enables the fractal embedding of energy flow.

At the center of the toroidal energy field is a black hole.[5, 6] Black holes are regions of space-time from which gravity is so strong that it prevents light from escaping. The infinity symbol is a cross section of the toroidal vortex. The midpoint of the symbol representing the black hole of contraction, the "still point" at the epicenter of infinity. The universe is an ecological system of renewal. In the energy flow of the toroidal energy field, first we have the gravitational collapse into a black hole, then the expansion, like aiming a jet of water into the sky, then gravity takes hold and the flow is reversed, making an oblong journey to the opposite pole. Then there is a contraction as the energy stream feels the pull of the black hole and is sucked back into it.[7] Then the whole cycle starts again. Expansion and contraction, creation and destruction, there is a constant renewal and regeneration of life. The black hole of contraction accompanied by the white star of expansion, constantly exploding. The cosmos expands and contracts at regular cycles as explained by Sri Krishna in the *Bhagavad Gita*; just like inhalation and exhalation.

Black holes vary in size, from the black hole at the center of the universe and the black hole at the center of our Milky Way galaxy, to the tiny black holes that exist within our own cells and energy centers. In our bodies, our hearts generate the biggest toroidal energy field.[8] At the center of the center of the heart's energy field is the largest black hole found in our body. Within this center, beyond time and space, there is a direct connection with "the all." Similarly, on a micro scale within the nucleus of atoms, the proton itself is a black hole.[9] It exists as an interface between the source of all, the void, and our physical manifest reality.

Just as each heart carries all the information of all hearts, all protons too are entangled and each proton carries all the information of all protons. Each one of us is directly connected to the entire universe, and can access all of the information within it at any given moment. This is the basic principle of the holographic model of reality, that all of the information present in the whole is also present in all parts of the whole. You contain the entire universe in each one of your atoms.[10]

The still point, the point of singularity, or in Sanskrit, the Bindu, lies at the center of a black hole. This point is both infinitely small and infinitely dense. It is a point that is beyond time and space. This point is within everything and it is your connection to all knowledge. It is the place where spirit washes in and matter takes a dip into the ocean of the spirit. It is where the inside becomes the outside and the smallest of the small opens to the expansive infinite. This place in the heart has been known by many different names in spiritual traditions. It has been called the "eye of the heart," "secret chamber of the heart," "tiny place within the heart" and others call it the "sacred chamber" or "the sacred heart." Whatever name you give it, when you can access this place in the heart, you are in a place of such stillness and peace, you can feel infinity.

Don't let it bother you if you can't logically understand the quantum world and the unified field. Even Nobel Prize winning physicist, Niels Bohr said "Anyone who is not shocked by quantum theory has not understood it."[11] Another of the world's leading quantum physicists was quoted as controversially saying, "If you think you understand quantum physics, you don't understand it."[12] Some quantum concepts like all information being available in all points of the field, and that something could be both infinitely small and infinitely dense, can baffle our brains. However, this information is perfectly comprehensible when felt with the heart. Feel this with your heart, don't try and think of it in a linear way.

Some knowledge must first be felt from within that place within the depths of the heart where the inside of the inside, the smallest of the small becomes the expansion, from that part of your heart that IS the field and knows the truth of things.

The vibration of your heart now beats in resonance with an unknown force. This force brings with it a sense of completeness. There is a feeling now that you are reaching deeply into a part of yourself that only knows one thing, and this is your true ability to trust your own power to create in your world your "Black Heart." For what is the Black Heart if it is not connected to the source of your primordial power. The raw primordial power will unleash an almighty force of energy, light, and power to you right now.

The Black Heart is an entry point of absoluteness in everything. The absoluteness of everything is in the Black Heart. It is the sum total of your absoluteness. Your absolute point of no return when everything is nothing, and the space of pure nothingness. Complete dissolution is creating in you the purity of your life in all its new forms.

Truly the forms of your life bring you to this space of purity and light for everything. The purity and light for everything in your world creates a space of this absoluteness for you. You must feel this absoluteness before anything can happen. We gravitate toward the black heart when we feel the absoluteness of every living thing. The sense of absoluteness takes you into a space of completeness; complete, the Black Heart completes a cycle in your life.

When you feel a cycle has completed, you rest in the Black Heart of the Mother. Always to return to the Black Heart is to remind you of your own ability to find within yourself what you need – the complete sense of pure nothingness. This sense of pure nothingness brings you to a space of absoluteness again. Feeling this sense of inner stillness now brings you to your own sense of complete and easeful gratitude for who you are.

When we interpret sacred texts and spiritual teachings with the knowledge that black holes are portals to the unified field and that these portals exist at the center of our atoms, at the center of our hearts, at the center of our chakras, our solar system, our galaxy, our universe and also that this unified field permeates all things, then there is a deeper, more experiential understanding of the ancient scriptures and spiritual texts.

> *"The first peace which is the most important, is that, which comes within the souls of people when they realize their relationship, their oneness with the universe and all its powers, and when they realize that at the center of the universe dwells the great spirit and that this center is really everywhere, it is within each of us."*
> ~Black Elk[13]

The sacred heart is a portal direct to source. When you enter the portals of your heart, you have access to the great wisdom and teachings that are there for you, and you can regain the dimensions of intelligence that cannot be learned by the mind. These dimensions of intelligence can be learned by experiencing the secrets of secrets, your heart's sacred chamber. Each time we surrender to our truth, more wisdom, more power, and greater love is revealed to us.

Thoth speaks of the knowledge held within the heart in the Emerald tablets.

"Desireth thou to know the deep hidden secret? Look in thy heart where the knowledge is bound. Know that in thee the secret is hidden, the source of all life, and the source of all death."

The Bindu, represents the point of singularity as the center of the ancient Buddhist symbol, the Sri Yantra. The Sri Yantra represents the cosmic unity and the underlying geometric energy structure of the vacuum that is a 64-tetrahedron grid. [14] Within this tetrahedron grid can be found many sacred symbols, the seed of life, the tree of life, and the six-pointed star. That 64 is the minimum number of tetrahedrons to begin to see the infinite fractal pattern is quite significant. There are also 64 codons in the double helix DNA of humans.[15] When a baby is first conceived, its cells divide in a tetrahedral pattern and each of the cells is identical until there are 64 cells, then after that they begin to differentiate into different body parts, one becomes a skin cell and another a bone cell and so on.[16]

If you place spheres around each of the 64 tetrahedrons that make up the structure of the unified field, the spheres overlap and are perfectly space filling, with no gaps. If you shine a light through such a structure, you see the Flower of Life symbol.[17] The ancients left us plenty of hints and the Flower of Life geometry features in artifacts found throughout the world. The temple of Osiris at Abydos, Egypt is thought to be the most ancient of the Flower of Life symbols found in the world today.

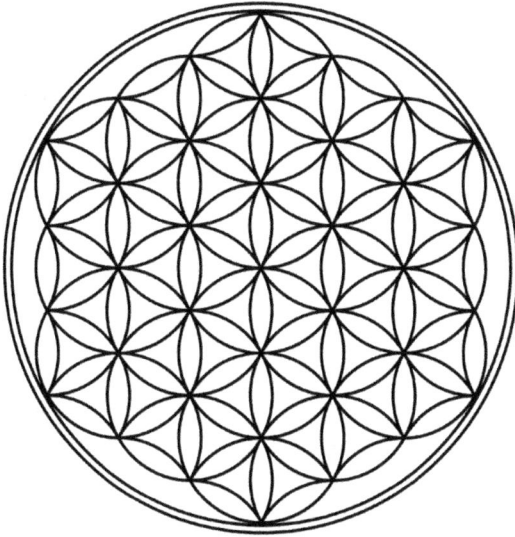

Ancient temples and places of knowledge exchange are not the only places where the Flower of Life symbol shows up. The Flower of Life structure is present within each and every atom. Protons are made up of subatomic particles called Planck spheres. These Planck spheres are thought to be the teeniest particles in our working universe. In a breakthrough, paradigm-changing discovery, it has been found that the spherical Planck's of the proton are configured in an interlocking Flower of Life pattern. Interfacing between the material world and the vacuum, the Flower of Life Planck configuration of the protons is reflecting the tetrahedral structure of the vacuum.[18]

The conscious knowledge of the Flower of Life configuration of the Planck spheres making up protons brings to humanity a new level of consciousness and a direct understanding of the connection to the undifferentiated source field. This knowledge has further opened the unified field to a humanity able to consciously resonate with the field of "the all" when you are able to bring your heart into resonance with the field, and consciously merge with it. It is a feeling of almost unimaginable love and feelings of ecstatic yet incredible peace and almost unbearable joy. There is an infinite expansion and within the surrendered sacred heart there is more love than ever thought possible. Poets and mystics throughout the ages have tried to convey this feeling, we simply don't have the words to fully describe it, and it is an experience that goes beyond verbal description. It is also the key to some wonderful advances in energy generation and space travel.

We now have the scientific quantification to align with the teachings of the masters. In the words of Thoth, *"Know I that man in his movement through space time shall ever be one with 'The All.'"* Connect with the truth of yourself, and go into that place of peace and stillness that exists on the inside of your heart and commune with the creator in your heart.

The presence of stillness is a palpable thing. It is just part of you, and this stillness is reflected everywhere. Life offers myriad possibilities when stillness is part of your world. Feeling this stillness now reinforces your self worth. For when the heart is still and silent, peace and order are created. Allow this feeling now to envelope you. This feeling of peace, love, and beauty brings to you all you need for your life of "All Love." When you are merging into a new consciousness, there can only be a sense of reverence, for all there is allows you now to partake in the mystery. To partake in the mystery of your heart's awakening is the greatest gift of your soul's intelligence.

Your soul's intelligence which is reflected and mirrored by your heart's intelligence can only be measured by the love you give and receive through your heart. This act of remembrance from your heart, allows you now to really feel the peace and light of all there is. The peace and light of all there is creates stillness and order. Stillness and order bring respect for all there is. All of life and you are one. You are the sum total of all life. All of life brings you to a birthing space, one of truth and mergence for all of life. Allowing this truth and mergence to resonate in your heart allows the special magic of who you are to bring you to a point of center in your heart where your heart is one with every living thing. For when you create with every living thing you become one with every living thing.

At the deepest level of your being you are one with all that is.

Allow yourself to explore this space not in a linear way, but just as a heartfelt knowing, that you have known it all along. Be kind to yourself in this process.

The field that you never left awaits you.

"Heart, I love you."

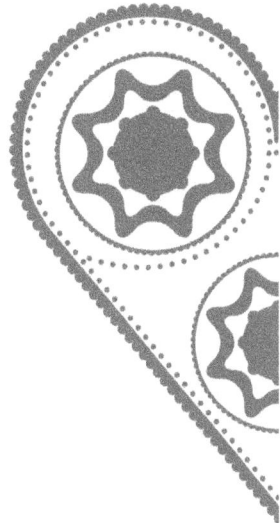

✑ Level 1 Activation ✑
Exercises for Awakening the Intelligent Heart
Self Nurturing Energy Belt – Base Chakra

To open our hearts for activation we must recognize that our own hearts need acknowledging. For the heart's intelligence to grow in human consciousness, the experience of nurturing is essential. Imagine now you are welcoming your heart into the world. Place it in your hands. You have just given birth to it. The feeling of being welcome is something our heart needs to experience daily or hourly if in crisis. The following exercises support your heart in its first initiation. Just being loved and truly adored by yourself is the greatest and all loving experience you can give yourself. Try it. For one hour, continually reinforce your "All Love" feeling for yourself.

Just continue to maintain the vigil for one hour every day. This is the surest and most complete way you can begin your journey to your heart, the undiscovered part of yourself that only knows one thing. The thing is "yourself." You are "the thing" itself, and this "self" reinforces total commitment to all there is.

You need to really feel and express to yourself your commitment to living through the heart of "All Love." The heart of "All Love" lives with peace and truth, and the peacefulness you experience must be above all else. Just allowing this precious peace to envelope you brings you home to the world of power and light. You are allowing this precious gift to envelope you and keep you in a space for all there is.

Bringing yourself now to the heart of your truth allows the energies through the "All Love" guides to begin to shape your destiny. Your truth through the "All Love" guides allows now for the state of bliss to be manifested in every living thing. The state of bliss brings, peace, tranquility, and order. Just feel this order now, as the great "she" brings to you all you need for your love-filled life. You are vibrating to this state of bliss now. Attuning to this energy only brings to you the precious magic of "All Love." The precious magic of "All Love" is now allowing "All Love" to infuse your very being, and your molecular structure is responding to this new way of being. Allowing this new way of being brings truth, aliveness, and magic to everything you feel, do, say, and attune to right now. This is a new reality and you need to express and feel this new reality shaping your destiny. Your destiny is a malleable thing. Your destiny brings to it a sense of wonder for all there is, and a sense of joy and truth for the precious gift of beingness is yours right now. Just now, you and your total beingness are one; just "one" for you and all of life.

Are you ready to commit to one hour every day, for one month, resting and reinforcing loving feelings toward your heart?

Ancient Egypt's Goddess Isis supports you now in the birth of your new heart. As you journey between the pylons in her holy home, the Isle of Philae, you arrive at her Mammisi, her birth room. Isis begins to form and remind you of her ability to cradle you in her heart as you bathe in her sacred waters. Your feeling of remembering your heart's truth begins to form. Feel this now. Breathe in lovingly now. When you begin to remember the power of your heart's capacity to feel, you are allowing the source of "All Love" to be your guide. This is a time to activate these forces in your remembering.

Activating these forces in your remembering brings you to a space of light and peace. Light and peace becomes encoded for your heart's truth and your heart's truth has this remembering encoded in it.

You are activating a dormant part of your remembering right now, as you bring to your world your heart's truth. You must begin to feel this heart's truth in everything as you allow the forces of "All Love" to bring you all you need for your journey. Feel now the precious gift of your remembering. Your remembering, your power to love, brings you home to yourself. Your "self" is the worshipful behavior of yourself. This is time now to create one-ness in this remembering.

Observe now, your own capacity to really remember when you had this much power and light. This power and light creates a vortex for more power and light. Allow the power and light to be your guide now as you bring to your world "All Love" and trust. The feeling of "All Love" and trust is around you now. Let's share in this collective knowing that we are all one in this remembering. All of life celebrates this remembering. All of life is one in this remembering.

By going into the conscious acknowledgement that your heart remembers, you can breathe love into your heart now and allow yourself peace and light. Feel your heart now open like a petal; next petal; next petal and so on, until your heart is fully opened to receive. Your heart cannot open, unless it receives. Allow the receiving to take place now as you bring to your life your own ability to know you have everything with you, to bring you all you need to feel. Just to feel, to really feel brings you into a space of truth for the mystery.

This super intelligence, your awakening intelligent heart, has just "received." The heart's primary receiving for you comes from humankind's sense of truth and purpose. The feeling of heart activation takes you deep into your core essence self. As this self is explored, there is a new vibration shifting through your molecular structure. With this shift comes a new awareness of yourself as being past, present, and future all in one. This shift brings with it a sense of power for the possibility of what you can achieve in your world.

Your world revolves in concentric circles, passing in and around in ever diminishing circles; concentric circles of power and intensity. This intensity creates in you a feeling of being in many different dimensions all at once, so it is important to remember you are human and you are part of the molecular structure of the Earth.

The Earth's molecular structure creates a state of connection with all there is as you bring home to yourself all you are capable of, for the new Earth shifts right now. Your energy just goes right into the new frequency where it begins to start re-creating itself and renewing itself. Feel the concentric balls now spiraling into new galaxies, new openings, and new portals. Right now you are bringing this resonance right into your molecular structure.

This activation has been a most life-changing experience for us all, and we must now honor ourselves as vehicles of incredible light and power. **These energies must be firmly anchored deeply into the core of the Earth right now.** You must just allow the source energy of "All Love" to bring you now to a space of pure completeness for all there is. This is a time for allowance and peace in the process of renewal through the moon's energies. What you have put out, you get back through the mother earth's heart and the moon's energy.

Talk to the Earth's heart now. This Earth heart takes all. The core heart of our Earth must be acknowledged now in our own hearts. The heart of the earth radiates such power to us now. Feel it, really feel it with your heart. Feel that now. Trust that what's happening will bring you this precious divine abundance of perfect health, emotionally, physically, spiritually, and mentally. By allowing your heart to tell you what it wants from you, you are being truthful and your relationship with the earth's heart is growing, bringing you even more earth power.

Right now you have had your heart opened for remembering, feeling your power and believing in it. You need now to release identification with your former self and know today you have located a capacity to really know and understand your heart's true remembering. You have to now really feel the energy of this power; your power. You have harnessed this power now, and it will change your life forever.

Intone:
"I have had my heart fully opened to access my true remembering."

Feel that now. Your truth is in remembering and you are now remembering your power.

Allow this new space to envelope you and bring you all you need for your journey. Feel that space within you now. Bring this space deep into your heart, as though you are painting a picture on your heart now. Get the pencils, the crayons, and oils and begin to paint this scene on your heart now. Do this now. Place the picture in your heart.

When you bring the space of forgiveness for yourself and every living thing, you are allowing this magic to happen. Allowing the magic of "All Love" to surround you brings you to a space of purity, life, and absolute power for all there is the sum total of the absoluteness of every living thing that has loved, and will ever love.

Just for now embody "All Love" as a part of your totality. What does it feel like? It can make you cry, you may feel you want to choke; you need to feel that you are free to embrace "All Love" and allow yourself the joy "All Love" brings. Just for now go deep into your heart. Really find a space that finds you vibrating to your own molecular structure to embrace "All Love" as a way of being human. Feel that now.

Feel deep within yourself "now" a space and place that only knows one thing. This "one" thing is the sum total of all your incarnations which have embodied "All Love." This sum total of all these incarnations now brings you to a sense of peace, space, and order in your life. Just begin to feel that now. How does it make you feel?

Hold the energy of love for yourself now, as you embrace this new view of yourself, just find within you, your new view of yourself. What are you seeing? Feel the light hold the experience of "All Love" as you allow your heart to bring you its truth. Right now you just need to feel that you are able to challenge all core beliefs around your identity. Your identity shapes your reality. Get clear about your "core" identity now. Really feel your "core" identity bringing you all you need for your journey.

These exercises have given you the opportunity to create a world where "self nurturing" begins to be encoded in everything you experience, because you have recognized your own capacity to love the "self." The "self," your heart, grows from this space. Feel this now.

The intelligent heart is activated when you acknowledge you need help in unburying your fears, the "All Love" guides are willing to assist in the recovery of your lost "self." Your lost "self" is in the wilderness and you, as a human, have created a form, a body to uncover these parts of yourself. Your relationship with your guides just keeps getting stronger and stronger, as you go into the "secret" self. Your secret self creates a space for this inner exploration to take place safely and with humility for what you will discover about yourself.

As your intelligent heart begins to awaken, it becomes necessary to balance yourself by opening your base chakra, the energy center at the base of your spine. This energy center resonates to the color red. Create your "Red room" and turn it into the most beautiful example of what you can imagine for this energy center.

Begin with creating your Red room now. Re-gut it. Decorate it and begin the construction of what you want this secure room to look like. Evoke the senses. Really create a sacred space within yourself to reclaim your lost identity in your Red room. Create it by visualization, touch, sound, taste, and smell.

What does your Red room look like? How does it feel? What are its textures? What sounds can you hear in there? Is there music playing? Or some other sound you can hear? What does it taste like? How does the air taste? The decorations? How does your Red room smell? What are its particular aromas?

List all the attributes you want for your Red room in your base chakra now. This exercise is very important for re-claiming your identity with yourself. Your identity is now being honored and treated with respect. Allow yourself to do this NOW. Bring all the energy centers into a state of resonance with your divine truth and you will have created a sound relationship with your own security. See this taking place within your consciousness now. I am alive to the sense of oneness, to the sound, to the touch, and the taste. All of myself is in celebration now as I come home to my base chakra's "Red Room."

Love and peace to you now as you have created this sacred center and make yourself secure. You will find there is a sense of peace, wholeness, and security in your relationship with this energy center. This energy center is now being prepared to strengthen you for your ascension. It is impossible to build a house without a firm foundation, because the structure just collapses to the ground. You must reinforce this "security" center every day; it is a center of great power. See it as a big red furnace, like a ship's furnace, which must be fueled with loads of energy daily, the fuel must be continually fed for it to stay alight. You need now to create this energetic space for your furnace and nurture it with quality fuel to keep it burning.

The fuel for your power center is breath and visualization. Visualization must be from the Earth. Much energy comes to this center from the Earth. The Earth is truth for this center. Visualize the red earth, the great red earth, creating fuel with your breath for this center daily. The daily task must be to fuel this center. Try and feel the love and energy creating power in you now. You need to feel the power, creating strength in the lower part of your body, just feel the power of your new creation. This is a time for surrender and trust in yourself as you journey inward. This is a time to create energy and power, as you bring to your life, your power to know you are capable of love and truth for yourself. Bring to your Red center your love and your heart. Bring your heart to your center now. Fill your Red room with something, try visualizing a red ruby. The journey upward will commence now, with the Red center fully secure. When your self-nurturing begins through this visualization, you can then begin a dialogue with your own heart.

In the name of Love and Light and in the name of my absolute divine truth, I now summon the "All Love" guides in their pure manifested form to be here and present now to bring light and order to the heart of…(Put your name in here). Now, ask your own heart: "Heart, what would you like to communicate for my soul's journey today?"

∾ Level 2 Activation ∾
Activating the Intelligent Heart's Endocrine Function
Self Respect Energy Belt–Belly and Solar Plexus Chakras

Now that your heart has been awakened, it is time to introduce it to its many functions. By initiating your heart to its endocrine function, you are now co-creating with yourself in wellness, bringing strength and rejuvenation to your newly awakened heart.

The heart is a flexible organ, a living, breathing intelligence, a huge hormonal pumping station. It responds to emotions like sadness and loss by pouring more energy into the troubled wound; if the heart goes into shock with grief and loss, the energetic damage can be enormous, as it puts a great strain not only on the heart, but the entire glandular system. Right now you can activate your heart's great reservoir of power through the moon for restoration, renewal, youth, and inner harmony.

"I am now activating my heart as an endocrine gland, the 'Master Healer' that controls all the glands of my body."

"I now ask that this awakening be taking place now in my cellular memory."

"I turn toward the moon and the restorative powers of the moon to renew and heal my heart."

"May this be taking place now in the name of love and light and in the name of my absolute divine truth."

Now the focus is on you, and you are in the spotlight. Just imagine you are formally in the spotlight in your life. Find your center. Do this now. Find within yourself your absolute divine center, and radiate out to the world, this center; the center of your heart's endocrine gland. Feel this now. Just allow this feeling to bring you all you need now. Right now you are the center of your life, and you are embodying that center, this center being your heart as an endocrine gland!

What is it you are embodying? Happiness, grief, loss, abandonment. Right now you are processing the enormity of being human: your cellular memory is now being recoded to accept new frequencies and new ways of being. You need to feel that you deserve this belief in yourself, and you need to feel now that you are one with the essence of all life when you embrace your heart as an endocrine gland. Believing in yourself is now one of the most challenging things you can do.

What stops you from believing in yourself? You need now to really feel this belief in yourself, in your absolute capacity to give yourself all you need for your love-filled journey. Feel this now. Allow the essence of this love to bring you all you need now to really heal core belief patterns about your deservability.

What is your deservability anyway? Your deservability is the essence of what your heart wants! *What does your heart want right now?* Peace; happiness; stillness. My heart now wants to embrace my heart, as an endocrine gland.

"I now embrace my heart's endocrine function, and am able to rejuvenate my whole body, mind, and spirit."

As you begin now to allow the truth to emerge your energy will shift, and **you will need to ground the incoming energies.** You need to trust this new space and give abundantly to yourself. This is now your time to really connect to the world of pure peace and magic for your life right now. Just trusting in this space brings you to a sense of knowing "you" and only "you" are responsible for all that is happening to "you" right now.

To really feel this "you" requires a special trust in just what "you" want for your life. What "you" are wanting for "your" life brings you home to your immortal heart. By attuning to your immortal heart you are just encoding yourself now for the truth to emerge.

The truth must emerge at some point in your journey in being human. You just have to face your oldest and biggest demon – your inability to be truthful to yourself. Why do humans find it so hard to be truthful to themselves? It is because they fear the knowing that the final betrayal is to themselves; and that they are their own greatest betrayer. You must continually reinforce your own truthfulness to yourself at every opportunity. *"Am I being truthful to myself right now?"*

Ask yourself: *"Am I being truthful to my heart right now?"* Asking your heart what is truthful; brings you to a space of remembering. Now I have to remember. This is tough for the heart, as it has lost its remembering button. Right now, you and I are going to activate the remembering button.

Ask: *"In the name of love and light and in the name of my absolute divine truth I now encode my intelligent heart for remembering my truth."* *"I now lovingly allow myself to remember my truth."*

Remembering your truth brings you deep into yourself where you will begin to activate old worn out memory patterns. These disused bits of fragmented DNA need rewiring to hook into your new reality. When you begin to do this you may like to reinforce "why" you want to live a truthful life. A truthful life allows your heart to shine. Your heart cannot grow or feel peace without knowing that "you" are truthful to yourself.

Getting to the core of your heart's remembering begins the activation to encode the cellular memory to bring you all you need for your life. Right now your consciousness has prepared you to embrace the new reality of your heart's endocrine function, and you can rejuvenate through it. This is truth for the heart. It remembers now. It has endocrine function! Remembering your truth is just that, "remembering," when we don't "remember," we bring chaos, fear, and abandonment. Challenge yourself now to remember your heart's endocrine function; you're remembering to live fully for you.

The joy of allowing the heart to bring you its magic is in allowing the truth of your heart to be revealed to yourself. *Why do you want an intelligent heart with full endocrine function? What is in it for you to want to embrace this truth?*

Your life force magnifies enormously when this energy is around you, and you must feel the magic, essence, and life force of all around you, bringing you all you need for your life. Just feeling this essence allows the sacred forces to create with you. The sacred forces correspond to your heart's intelligence as they are attuning to your heart all the time.

All attunements to the heart take place in the company of mighty intelligences that are sent to support the soul in its evolvement. This evolvement shapes the destiny of all the forces in the universe as well, and your hearts intelligence grows with all the remembering you are capable of. Believing in this magical capacity allows the forces of love and nature to combine to bring in magical intent. Bringing in magical intent reinforces the soul's journey to feel and remember all there ever was, and all there ever will be. It is time now to allow this special magic to bring to you all you need for you life of sweet remembering. The past, present, and future merge into "one." "One" knowing, "one" love, "one" heart. For we are all part of the one heart anyway.

"My fully functioning intelligent heart as an endocrine gland brings the mighty intelligences to me supporting my soul in evolvement."

When we love ourselves enough to experience this energy, much, much more is being produced. Ask yourself now, how does that feel? Describe how you feel right now being able to access this incredible download of raw energy to yourself. Activating the heart's endocrine function brings to the intelligent heart a space to experience bliss! When you are bliss filled, endorphins fill the body with light and energy.

You are releasing these endorphins into your body now, charging it and filling it with light. Allow this process to take place now, so you can fully embrace the light. Feel the power surge through your body, opening and charging it to receive even more light and power. When you allow yourself to experience this energy, more energy is produced and you are holding light, power, and strength in your body.

Imagine pressing a button in your heart center now and filling it with light and energy through fully functioning endocrine hormones. To reconnect your own individual endocrine system to full function has far reaching benefits for both your individual well being and the well being of the collective whole.

The heart's endocrine function is a fully intelligent molecular structure that responds to thought through vibration. The vibration through thought awakens it, and its recoding is dependent upon the food you feed it. When was the last time you said?

"Heart, I love you!"

The enormity of such a simple statement addressing your heart can be profoundly moving and it is possible to be so moved as to spontaneously burst into tears, just by the simple statement:

"Heart, I love you!"

When you feed your heart love, it responds by remembering. It is awakening to this remembering now. Express this to your heart daily and in troubled times, hourly.

The ancients knew of the interconnectedness of the universe and in many ancient and indigenous cultures there has existed a practice of connecting your heart with Mother Earth and Father Sky. You can experience this practice, too. Start by breathing deeply and relaxing. As you are breathing deeply and relaxing, feel the deep love you have in your heart, keep your consciousness in your heart, and just breathe and connect with the amazing depth of love that exists there. Now, feel the love that you have for the Earth, this beautiful planet that supports your living, feel your connection to her being, to the heart of the Earth, and consciously send the love that you have in your heart for the Earth to the Earth's heart, deep into her core.

When you feel that love has been received, feel the love that the Earth has for you, feel that love come into your heart, continue the relaxed deep breaths as you enjoy the heart connection. When you are ready, feel in your heart the love that you have for the cosmos, for the stars, and the sun and consciously send the love you are feeling in your heart, to the heart of the great central sun, and then to the hearts of "All Love"-filled star constellation civilizations. Hold the intent that the stars, the sun, and the Earth are co creating with you right now for this interconnectedness to take place. When you feel your love has been received, feel now the love that the universal heart and star beings have for you as their love comes into your heart. Sit for a while; relaxed and breathing deeply in this field of love.

Now, if you like, feel eight rings of golden light circling you and bringing your heart into resonance with your truth. You are feeling your power and light to bring you all you need for the next stage of your journey. The essence of your heart shines and you are feeling the magic and the light all around you. Right now your must see your heart being bathed in iridescent golden light as you draw down this extraordinary energy and light.

To be one with this energy, at this time brings you into a pure space of remembering all you ever were. Right now remembering all you ever were brings you into complete alignment with your truth and power.

For now, you are a witness to your own extraordinary process in being human. This process is taking you back to a time and place where all you ever were, will be activated. Allowing this activation right now brings you into resonance with the whole stellar heart that supports this emergence. Your emergence into this stellar heart now brings you closer, to the group stellar heart of "All Love." Attune now to the stellar heart of "All Love." Right now your heart is being activated for stellar heart energy and light to come to you. Breathe in the stellar heart now. The stellar heart is now being encoded into your cellular memory right now. This energy brings peace and power. Right now you must feel the energy of this resonance bringing you all you need for your world of remembering one thing, your power to love. You are love...

"Heart, I love you!"

This next powerful ritual can be conducted physically on a beach or in a space where you can see the sun rise and simultaneously at dawn, feel the power of the moon, which can often be directly parallel to the sun in the western sky. If you cannot do this in nature when the moon is full, just develop a ritual anywhere in nature at dawn when the sun is beginning to make its ascent in the eastern sky.

Facing the sun, open your heart to the sun, really allowing the powerful life-changing beams to deeply penetrate your heart's core matrix, its essence, energizing and enlivening the fluids of the heart.

"Open your heart to the sun calling upon the vigor, light, and masculine strength of it, to balance and open your heart's endocrine function."

Create this intent, really feeling it now. Allow this intent to balance the male side of your heart, its right side, for energy, light, and power. Feel this now.

When you feel this process is complete, you face west and see or visualize the moon bringing in the restorative, regenerative powers. Really feel the moon's light and power bring you into internal resonance on your feminine side. When this process feels complete and when you have felt the moon's power deeply penetrate your heart's core matrix, allow yourself time to feel the moon's magnetic energy deeply merge into your heart's core essence.

With the heart sun power activation and the moon's regenerative power, bring your left side and right side of your heart into resonance with each other by visualizing an infinity symbol distributing the magnetic restorative power of the moon's energy and the enlivening manifesting energy of the sun. Feel this infinity symbol all over your heart taking in this elixir, ambrosia of life, bringing to your heart, the strength and power of the sun and magnetizing the restorative energy of the moon simultaneously.

You are in a space of true reverence for this alchemical process.

Now deeply, powerfully, and rhythmically breathe into your heart its essence, its truth, and its power to bring you all you need for your journey. The combination of peace and vigor brings you deep into a space within yourself, which knows exactly what you must do with your newly awakened intelligent heart.

Honor yourself now in this intelligence.

Ask yourself:
Is my body functioning to its fullest capacity? Yes / No

Is my mind functioning in a way that reflects my truth? Yes / No

Is my spirit fully alive and awake by me living fully in the present moment and not distracted? Yes / No

To balance and bring in this new intelligence intone – "Heart, I love you," focusing on your body being perfectly healthy, your mind functioning in a way that reflects your truth, and your spirit being fully alive and being fully present in the moment.

◈ Level 3 Activation ◈
Activating the Intelligent Heart's Secrets of the Feminine DNA
The Self Energy Belt – Heart and Thymus Chakra

One of science's most badly kept secrets is the incredible power of MtDNA. Your consciousness is now entrained to begin to discover the source of The Self. The secret is yours now.

The energy of the MtDNA is now available to begin the journey to discover the source of "the self." The source of "the self" is the sole purpose for humanity's evolution. Humans cannot evolve as a species unless there is an activation of the MtDNA as there is no glandular stimulation to love. Primarily the sole purpose for humanity's evolution is contained in this DNA. If humans choose not to evolve while in a human body, this gland remains dormant and in a passive state.

Right now the energy of "All Love" activates the gland, which is primarily located in the heart. Bringing this energy into resonance with your heart's truth is awakening the heart, releasing it from its slumber, its inertia.

When you consciously activate the MtDNA, you are stepping into a new world of multidimensionality. Suddenly linear time does not exist in the way if once did. You have tapped into a very rich resource of power and light. This power and light activates all cells in your body. By activating the MtDNA, all DNA becomes charged and activated as well.

The true essence of rejuvenation takes place. The cells can be activated to awaken dead hearts not only energetically but also physically. You need now to feel your source energy of "All Love" bringing you all you need for your power to heal, rejuvenate, and bring source energy of "All Love" to you now.

This transformative energy is awaiting you.

Are you ready to receive the power of MtDNA activation?

"I now allow myself to receive power and light through MtDNA activation, bringing the true essence of rejuvenation."

MtDNA brings to the surface an undiscovered source of raw power. This DNA activation is like tapping into a rich reservoir of raw power. This raw power brings so much energy to the whole nervous system. This is very important to keep focusing on just how magical and life changing your world is becoming with the source energy of this DNA.

Just focusing on its source power will give your heart strength. The heart is flooded with energy and power, as the gland becomes activated. The energy of this activation brings to the surface much suppressed grief as you begin to transit into a new view of your reality and potential.

You can begin to feel the power of it as being similar to an atomic energy, exploding power and light throughout your whole body. Feeling your body become charged up with source energy and power, allows you to really feel the magic of "All Love" in your life. Your world becomes one with the source energy of raw power for your body. Allowing the feeling brings you closer to the power of the ancients who knew this DNA and used it to bring might and power to their world.

With the MtDNA activation, you will begin to experience a worldview, which encompasses light and power. Energy you did not believe you had will be made available to you. People who perform heroic deeds; i.e. saving others, etc. have had this surge of power to fight for the lives of others in a humane way activated. Because of this process, your whole DNA is being recoded to accept more light and power to charge your whole body. The body becomes immune to diseases that plague humanity because there is oxygen and light in all organs and tissues and the endocrine system become regulated. You become a true light body. Every component of your body is then activated for remembering your true power and light. To allow this process to take place brings you into alignment with your true purpose on Earth. Your true purpose on Earth is to find your heart's remembering.

Feel now your heart being bathed in golden light. The energy and power of golden light keeps you enflamed. Your heart begins to burn brightly. You have seen many representations of this flaming heart in depictions of your religions. Saints, holy people, and prophets have flaming hearts.

Activation of the MtDNA is a gift to you for rejuvenation and longevity and your heroic deeds are now encoded in you for fighting for your heart and the heart's truth of others. In the film "The Wizard of Oz," Dorothy's three friends each had to fight for their own remembering to face their inner demons, and also to fight hard for each other, as a group for this remembering.

Beginning the journey to activate the heart's intelligence brings to your life a sense of peace and order. There is a quiet reverence for all of life and an acceptance of what life brings. Your cells right now are undergoing transformation for renewal and regeneration. Rejuvenate your own DNA now by consciously activating the MtDNA by intoning:

"I want the whole body flushed from toxic material and that all DNA can be repaired when toxic material is released from your system."

Asking for this gift brings you into harmony with your own true purpose in being human. You will begin to live humanely and humanly for yourself. This process brings up suppression. Suppressed emotions that you have stored all of your life surface when the full component of your DNA becomes activated through the MtDNA. You are vibrating now to the release of toxins in your body, this process requires you to be still; **very still and peaceful in the process of cellular regeneration**. Regeneration requires stillness and peace. Remembering your perfect self now begins to surface, as your full potential is becoming visible.

The MtDNA is the trigger for the complete restructuring of your humanness. The definition of being human is now being offered to you as you demand to have "all the information" that being human offers.

Being human offers gifts way beyond your imagining. When your heart's intelligence is fully activated and awakened, this remembering becomes possible.

Allow the spirit of complete trust in the surrender process to bring your heart into alignment with its true purpose; it shines and begins to feel welcomed in the human journey. This is a sacred time in your Earth journey right now as you bring the peace and remembering to all that you do, think, and feel. The essence of all you do lies deep within you where you become totally one with all around you. This remembering of who you are reinforces in you peace, light, and order. This is a time to pour the activation of the MtDNA's throughout your whole molecular structure, even your skeletal system and marrow.

To feel this energy, brings light and power to you right now as you begin to really create a space for love and trust for yourself now. All around you is the opportunity to love and trust. The opportunity to love and trust brings you a new identity. This new identity allows you to go deeply within yourself discovering the lost treasures of your past remembering.

All around you is the capacity to love. Look at the things to love. Start saying to yourself that every day you look for experiences which reinforce the capacity to love in your life. Feel this now, and as you do, open up wider to the part of yourself that has love encoded in it. Love is encoded in every living thing. Finding that capacity within yourself to embrace your own capacity to love only what is loving, not what is dangerous or evil. Draw out the danger of your own fears now to love every living aspect of yourself. No harm can really touch you. Mostly these fears come from fear of your feminine remembering.

Reflect on the nature of the "feminine" as this is where the fear lay. The word "feminine" creates different reactions in people. Fear will be found in the resistance the masculine MIND puts up against the "feminine" remembering. By work shopping what "the feminine" means to you, you will begin to allow yourself to create a view of your world through the lens of the feminine heart.

Meditation to complement the activation of the MtDNA

Right now, just allow the surrender to take place within you. Feel the gentle surrender. You are floating on the Nile; the river of life and your felucca is just drifting in the wind. The breeze is gentle, soft against your skin. Feel this now.

Again, dip your hands into the silky texture of the Nile, its coolness reminding you to really feel peace in your heart. Allow the surrender to take place in observing the sand mountains, feeling energetically loved by all of nature. Feel this now, and just allow yourself to drift off in your own world, your own dreamtime.

To begin the journey to heal fragments of DNA and restore missing codings requires patience and discipline. The journey to the heart is fraught with perils as old patterns of fear keep resurfacing at a time when you are beginning to really feel your heart. You are forgiving a part of yourself now that keeps trying to destabilize new opportunities. You must feel now these new opportunities as challenges to destroy all that is not light filled in your life. You must feel the freedom now of mergence and light as you must honor the part of yourself that knows this truth and wants to live with total integration of all your energetic selves.

You are always going to have to challenge these belief patterns as you give your heart to yourself. For the gift of your heart to yourself keeps you in the remembering space, and you must constantly strive to remember your own light and power. Concentrated light burns. It burns away everything that is not needed in your life. Feeling this concentrated light brings you into total resonance with all there is, as you bring to your world peace, light, and power.

To achieve the delicate balance of peace, light, and power requires you to keep remembering all you ever were and all you ever will be. This is time for powerful cellular transformation right now. You are in a pure space for remembering now.

This process brings the power of the force; the statement "May the Force be with you" becomes a living energy for the MtDNA to be activated.

When we consciously activate our MtDNA, it is a gift of rejuvenation.

⤳ Level 4 Activation ⤳
Activating the Intelligent Heart by Awakening the Primary Senses
The Self Realization Energy Belt - Throat and Third Eye Chakras

When we allow ourselves to resonate with our senses, we begin to open to the birth of our hearts at a new level of consciousness. If we "intend" that our "hearts'" primary senses can be activated for "remembering," our intent shapes our destiny.

Allowing the forces of nature to assist in your heart's remembering brings you to a space of pure delight in the process of being human. You are feeling this light and power resonate through you as you open up to the world of a most spectacular light show in your consciousness.

You are asking now for this energy to be around your whole molecular structure, as you bring to your world this remembering. Believing in this precious remembering is allowing the space of light and magic to envelope you as you create in this remembering your power and light. You are feeling the energy surround you now, and you are magnifying your intent every time you give love and energy to your intent. For it is the intent that brings you home to yourself, and it is your intent that shapes your destiny. Your destiny is a malleable thing. Intent shapes it. Your intent can create miracles. Your intent creates your world. Intent is the miracle. Your activation of intent starts with a feeling of love for yourself, which starts out with one tiny strand of knowing and grows to a forest.

Imagine your intent magnifying to create a forest of unbelievable "self love." This forest of unbelievable "self love" magnifies and brings to your world "All Love," and "All Love" brings the whole essence of being into one whole. Allowing this essence to shape your world brings you into resonance with all there is. You are love.

The world of nature assists in the heart's remembering because it activates our five primary senses. When our heart's five primary senses are activated we begin to remember.

Your heart's receptors are very sensitive to the five primary senses and we awaken our heart's intelligence by activating those senses. Each person has five energy belts around the body. Each of the five energy belts relate to a particular sense. To activate the heart's intelligence you need to see the heart as having receptors which are very sensitive to the five primary senses of the energy belts. Each person has one of the five energy belts that the heart responds to as its dominant primary sense. The correspondences between the sensory perceptions and the energy belts discussed below are a part of the Philosophy of the Divine Feminine and are in relation to the senses of the heart.

The energy belt located at our *Base Chakra* area is the *"Self Nurturing"* energy belt. It connects to the pelvic area of the body. It is activated by *Touch,* and is the most basic primal connection to being in our body.

Our *Belly and Solar Plexus Chakra* is the *"Self Respect"* energy belt. It responds to *Smell,* i.e. I smell a rat, as a metaphor. It responds to oils, perfumes, food, and other aromas. All smells activate the core center, the belly of our being, our center.

The *Heart and Thymus Chakras* are *"The Seat of the Self"* and relate to the *"Self"* energy belt. The Heart center responds to the sense of *Taste*. When we feel through our heart with feelings of compassion, giving gratitude and appreciation, we open up the heart to remembering our true power.

The *Throat and Third Eye Chakras,* including the face and ears is the *"Self Realization"* energy belt. This is activated when we begin to feel through *Hearing* and listening with attention to the sounds of nature, music, chanting, poetry, and other inspired speech.

The *Crown Chakra and Beyond* is the *"Bliss"* energy belt and is activated when we *See* all; everything becomes clear and we can really see and live our truth.

As you begin to activate your senses through your heart, your heart will begin to "feel" safe and create with you. There is an outpouring of suppressed toxic material and you may wish to support yourself professionally in getting counseling, etc. for your pain. An animal caught in a trap cries loudly in pain, so do you, as suddenly the shock of suppressed emotion can find you stuck and hurt. Be gentle with yourself, but get the pain source removed. The animal must wait for release, or it can die. Be prepared to answer your pain, by getting help at many levels. Listen to the pain and act, release the source of that pain.

This pain is the trigger for your newly merging self as you drive yourself deeper into your heart. The feeling of knowing your heart is the very organ being punished brings you to a space of surrender, absolute surrender and faith in living primarily through your intelligent heart.

Your intelligent heart speaks volumes to you. Listen to it as you truly begin to ascend, taking apart all the pain it has endured and releasing this pain. Feeling this pain, releasing it, and moving into a new way of being human, brings you now a sense of ownership of all you have created. Your heart has time now to find this pain and open up to releasing, to its true source to love. Just to love.

Activating the hearts primary senses, one by one, stops the self-punishment of your very own heart. You wouldn't punish a child, pet or elderly person, but you do punish you own heart when you don't address your heart's capacity to create with the senses.

Activating the hearts senses supports this release. Breath will support this mergence of your heart's senses by creating a magnetic shield between you and your fully awakened, fully intelligent heart. Feel now the magnetic sheath be opened for you to bring this magnetic shield to you. For when you truly embrace this magnetic shield you are bringing to you the magic of your divine presence to yourself.

This divine presence is allowing you now to begin the mystery of your own creation. Keep the breath going, evoking through the breath your ability to hold light and energy through this magic shield. Breathe in the magnetic shield. Feeling it now protecting the opening of your newly emerging awakened heart's five primary senses. Your heart will begin to activate its own memory, opening now to the magnetic force of energy around it.

What are you creating?

I am creating a vortex of raw transformative power right now to bring me all I need for my heart's intelligence. Right now to activate my heart's primary senses. Feel your own heart calling to the sound of its own mystery. Be in this mystery now. For it is your heart which is the mystery and it has its story. As the story unfolds, so the mystery is revealed.

What mystery are you revealing to yourself right now?

"I am revealing the mystery of my heart's capacity to be awakened through its five primary senses."

Activating your intelligent heart allows you to really respond to your senses and respect them as living energies supporting you on your journey right now.

Giving yourself the opportunity to be a witness to your heart's mystery, a new sense of direction and purpose in her life, giving her power to recognize this power for the first time.

Begin to experience your power as an inner heart-centered energy right now by really allowing you and your senses to become "friends."

Say lovingly aloud:

"My heart's eyes…I love you. I see with you right now."

"My heart's ears…I love you. I am listening to you right now."

"My heart's nose… I love you. I am smelling with you right now."

"My heart's tongue… I love you. I am tasting with you right now."

"My heart's skin… I love you. I am feeling with you right now."

This new identity through your heart's senses activation needs support.

Ask yourself right now:
"Who is supporting my new identity right now?"

More importantly, ask yourself:
"Can I support my new identity?"

When you feel ready to support your new identity you will begin to draw new people, events, and situations to you.

The first support must come from you.

Breath gets you there. Begin breathing in; holding; releasing; pause. As you breathe in; ask yourself:

"Can I support my new identity?"

You need to give yourself love, peace, and tranquility to allow yourself to support your new identity.

"I now allow myself support in my emerging new identity."

"I am Love."

Allowing the source energy of "All Love" to be your guide brings you into a pure space of remembering all you ever were and all you ever will become.

This is a time to really allow the spirit of your newly awakening emerging spirit through the heart's primary senses to begin to feel and remember this "Bliss." Remembering this "Bliss" brings you into resonance with all there is. All there is, is the sum total of every living thing. Every living thing in your life has a parallel reality creating itself on another dimension in time. You are remembering this now. So who are you? This definition of yourself is mirrored to you in another time and place. This definition takes care of itself as you bring to your world all you ever were, and all you ever will become. Your relationship with this reality determines you earthly existence. Feel this now.

My earthly existence is determined by who I am.

"Who am I?" "I am a ..."

"I reflect to myself what I give out. I now give out to all universes, my capacity to create a world of pure peace and love for the mystery of myself, which is a multidimensional being having an experience in a human body."

If you intone every day: *"I now ask that my heart's intelligence be fully awakened,"* you will begin to bring to your world this love, and will begin to allow the sense of pure magic and trust to envelope you. This is a trusting process. We can perform miracles every day if we open our fully intelligent awakened heart to the source of all love.

Witnessing the birth of your intelligent, fully awakened heart brings with it a huge magnetic resistance. Expect some big changes as the forces of gravity resist in your body. When you are opening to these new energies, you will begin to find old patterns resurfacing and you will begin now to acknowledge that you often feel afraid, lonely, and abandoned. For it is the old pattern which has resurfaced at a time of great personal evolvement which needs to be swept away.

You may feel as though there is a "tug of war" going on in your body, as old patterns which you "think" you have finally finished with keep resurfacing like scum on a crystal clear pond. Where is "this" coming from? You may ask, I have dealt with "this" long ago. What is it in me that keeps repeating the same pattern over and over again? You must now outline to yourself what it is that you have not dealt with in your life. Again, you feel you must sift through the same old debris. You must feel this magnetic pull, like the Earth's gravity resisting your spacecraft flying through the air and into space. The density of Earth's gravity keeps you stuck in a pattern of old worn-out belief patterns.

Release this forever by saying:
"No, I am not this person. I am not you. I am now Me, I am only my Heart, and you now need to go."

This time you have had with me is over. Sweep away this old pattern so it is now completely obliterated.

Release your need to see this pattern creating any problems for you. The heart's intelligence is a force of almighty power, and the resistance in the body, mind, and emotions will put up a huge fight. You are now free to live fully in the heart, with all love and peace for a truly unified life. You are opening up to raw light power to transform your light body, as your remembering takes you deep within yourself.

Feel this extraordinary capacity to feel, heal, and remember all you ever were. Bringing this remembering into consciousness now requires an enormous amount of patience with the part of yourself that is fearful, lost, and abandoned.

Feeling this energy aligns you now to the special magic of your true remembering, of all there ever was. Finding this remembering in yourself now just allows for the magic to happen in your life. The magic of this happening brings you to a lost part of yourself now, as you struggle within yourself to really find this lost capacity. Within the light body there is enormous struggle to find this light switch, this remembering of all you ever were. Finding this remembering and bringing it into consciousness helps you stop feeling you are not part of the wonder, all creation is capable of.

Right now, just feeling this capacity to remember is allowing you to connect to the forces of "All Love" for your own heart. Observing yourself in your stillness shows you where your patterns become stuck in expression. For the heart, you must express your power to really feel love for each other. Bringing to your heart the magic of "All Love" reinforces the mystery of all around you. This essence allows you to really connect to the one heart, the heart of immortality. Just feeling the intelligence of every created being brings you into resonance with every living thing.

The essence of every living thing allows for oneness. Oneness is the state of "being" perceived by you in relation to every living thing. This state of oneness gathers momentum and draws to you a special oneness with everything in your universe that wants to participate in your growth.

As your heart is an intelligent organ, it must be opened wide to receiving information from many different sources around it. It needs many different experiences and states of evolvement, in bringing to it a definition of itself. For example, your intelligent heart only wants to see, hear, taste, touch, smell, and be part of a truthful existence. When this state of truth is not resonating there is a sense of abandonment and loss. This state brings to it atrophy. It begins to become stagnant, and the energy doesn't circulate through it. It is very important to feel the magical essence in your heart now as you awaken to the mystery.

The essence of all life is with you as you really open up to who you are. You are witnessing a new aspect of yourself merging now and you must feel the release that comes with this new identity. By allowing yourself to open up your heart to yourself you are beginning to remember, and in remembering who you are, you are also in the process of release and forgiveness too.

The balance must be maintained at all times. It is a delicate balance, for when we open our hearts to ourselves, we are also grieving. The old self must be released and moved on. There must be an acknowledgment of this and a respect for the process as well. You are now feeling that the time for loving the "self" brings you to a state of bliss in the remembering process. Remembering who you are brings you home, and you must now allow yourself space and love to allow what you see to be activated in your cellular memory. The activation of the heart's intelligence to bring to who you are; this allows yourself now the sense of joy and peace your journey brings you.

You are coming home to this part of yourself now. You are feeling now that the energy of pure love is around you and you need to bring to your world this love of self. You are witnessing a part of yourself that must be born, and this birth is reminding you that you are allowing this essence of yourself to emerge. Just now, feel the tiny seed that was planted before your creation; come into your human body, bringing it and your new life. The new life now is being pushed to the surface in a volcanic explosion of pure love. Right now you are birthing this new consciousness. It is you, birthing yourself to really feel and remember. For in the remembering you are being reminded to feel exactly who you are.

Who am I? I have a genetic blue print that connects me to my ancestors, my people, my race, and my heritage. However, I am more than this genetic blueprint. I have an energetic blue print which connects me to my past ancient soul ancestors. The ancients are our teachers and we become their teachers as well as we now co-create in oneness for all of life. These ancients are with you right now. They are just on a different vibration, vibrating to a different frequency.

Bring now your birth. You are now birthing your heart to its new consciousness. Your new heart is being born right now and your new heart is connecting you to the universal heart. The universal heart then connects you to "All Love," so you are spiraling upward in a cyclical dance of creation.

This creation is your "self," your heart. I am "All Love" now. I am my heart now.

⊗ Level 5 Activation ⊗
Activating the Intelligent Heart
Creating from the "Black Heart"
Bliss Center Energy Belt - Crown Chakra and Beyond

The five initiations to the Black Heart are surrenders; surrenders to dissolve, merge, and become "one" with our own "Black Heart." The Goddess void.

For what are we if we cannot find this space in our hearts on this human journey. Being consciously aware of this vibration is the sum total of your absoluteness.

The vibration of your heart now beats in resonance with an unknown force. This force brings with it a sense of completeness. There is a feeling now that you are reaching deeply into a part of yourself that only knows one thing, and this is your true ability to trust your own power to create in your world your "Black Heart." For what is the "Black Heart" if it is not connected to the source of your raw, primordial power. This raw, primordial power will unleash an almighty force of energy, light, and power to you now. The "Black Heart" is a point, an entry point of absoluteness in everything.

The absoluteness of everything is in the "Black Heart." It is the sum total of your absoluteness; your absolute point of no return when everything is nothing, and the space of pure nothingness, complete dissolution is creating in you the purity of your life in all its new forms. Truly the forms of your life bring you to this space of purity and light for everything. The purity and light for everything in your world creates a space of this absoluteness, for you must feel this absoluteness before anything can happen.

We gravitate toward the "Black Heart" when we feel the absoluteness of every living thing. A sense of absoluteness takes you into a space of completeness. The "Black Heart" completes a cycle in your life. When you feel a cycle has completed, you rest in the "Black Heart" of the Mother. Always to return to the "Black Heart" is to remind you of your own ability to really find within yourself what you need; the complete sense of pure nothingness. This sense of pure nothingness brings you to a state of absoluteness again. Feeling this sense of inner stillness now brings you to your own sense of complete and peaceful gratitude for who you are.

There is a sense of simple peace when you discover this new "Black Heart" of your human remembering. Your quest for peace in your life only leads you to a space of purity for all there is. Purity implies just that; "Pure." A pure heart is one not contaminated be fear. There is no fear where there is a pure heart. The joy of seeing someone with a pure heart is in knowing that they carry no fear at all in their energetic makeup. You need to really focus now on allowing this sense of absolute peace in your life bringing you all you need for your love-filled life. Just believing in the absence of fear in your pure heart, allows it to speak to you in a special way. This power grows in you now.

This power to remember; to remember a time when there was no fear, brings you to a special space in your remembering. This time now brings you to a space of "All Love" and "All Peace" for "All" there is. All of life and you are one. You and all of life are now one in remembering. This is a time to bring to your world this energy, joy, and love for all there is. You are part of this remembering now.

Allowing the truth to emerge brings you back to yourself. Bringing you home to you, allows you to bathe in the energy of you, for it is you who are one with all of life.

All of life and you are one. All of life creates with you now. You are the sum total of every living thing, and every living thing creates with you now in this spirit of remembering.

The presence of stillness is a palpable thing. It is just part of you and this stillness is reflected everywhere. Life offers myriad possibilities when stillness is part of your world. Feeling this stillness now reinforces your self worth. For when the heart is still and silent, peace and order are created. Allow this feeling now to envelope you. This feeling of real love and beauty brings to you all you need for your life of "All Love."

When you are merging into a new consciousness, there can only be a sense of reverence for all there is. This sense of reverence for all there is allows you now to partake in the mystery. To partake in the mystery of your heart's awakening is the greatest gift of your soul's intelligence. Your soul's intelligence, which is reflected and mirrored by your heart's intelligence, can only be measured by the love you give and receive through your heart. This act of remembrance for your heart, allows you now to really feel peace and light of all there is. The peace and light of all there is, creates stillness and order. Stillness and order brings respect for all there is. All of life and you are one.

You are the sum total of all of life. All of life brings you to a birthing space, one of truth and mergence with all of life. Allowing this truth and mergence to resonate in your heart allows the special magic of who you are to bring you to a point of center in your heart where your heart is one with every living thing: For when you create with every living thing, you become "One" with every living thing.

Right now birthing the heart is the greatest act of courage any human can give themselves, as it is making a statement about your right to live on this planet in such a way that reflects truth. This is a time now to put aside any fears about limitation in your life, as you carve out your truth. Your truth is a weapon for you to carve and your truth brings with it power and light.

Power and light bring the sense of magical renewal for all of life. All of life respects you now, as you bring to your world this act of remembrance. Just feeling now this act of remembrance allows you to really focus on the light of your heart growing for all there is. The light of your heart grows for all to see. Allowing this light to merge with you now feeds even more light and power to your heart. Bringing this light and power to your heart allows you to really own the part of yourself that has been lost, disconnected, and shut off from remembering. You are allowing the power and essence to bring you all you need to feel, see, taste, touch, and hear your truth. For when you activate the heart's senses you have a fully intelligent, fully awakened heart to love. Consider this now: You have a blue print for remembering your power, your light, and your peace because you have found it within yourself. You have unlocked all the doors.

Congratulations! You are now free to be "YOU!" A fully Awakened Intelligent Heart!

PART II:
THE IMMORTAL WOMAN

The Immortal Woman teachings go beyond time and place to embrace the fusion of the heart's intelligence. The Immortal's language reaches to a part of ourselves that knows how to feel; it strips away the illusion that the mind has "control" over outcomes. This section of the book has been written through the language of the heart.

When you read the words on the pages of The Immortal Woman *your heart begins to awaken to your soul's purpose through your own heart's intelligence, birthing a part of yourself that has been dormant. These are magical transmissions. It truly awakens the soul to the mystery of being human and makes you feel safe in being "human."*

The Immortal Woman *can be read sequentially or by opening up to any page for her message to be reflected upon for your day and see what "She" reveals to you for your journey right now. Close your eyes, take a deep breath in, ask your question, and then open the book up to a page. This is your personal "living guide" as the ancient secrets are revealed to you day by day.*

Profound wisdom is contained in every sentence as it speaks from the heart through the vibration of love. To become an Immortal enables you to experience a space of pure love and create in your life without emotional attachment. The Immortal Woman *tests and challenges your previously held definition of Love and takes you to a space of truth for yourself where all of life becomes a multidimensional experience.*

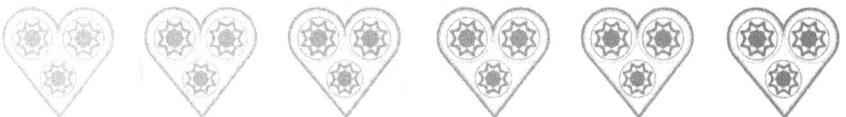

Chapter 6

SELF NURTURING

The fear of being human is to feel out of control emotionally, financially, or spiritually. Allow yourself control by evoking the Source of All Light itself—the Divine Feminine—to co-create within you a sense of immortality. To be immortal is to really feel the joy of the present moment—the absolute and precious moment of being alive now—and your future is being lived out in the present moment. The present-moment life is the key to the concept of the Immortal Woman. Humans are conditioned to believe in planning for the future and to believe that everything must revolve around these plans; this is aging. Doing so is limiting, as it creates stress and even despair.

Try intoning daily, even hourly, your personal mantra or affirmation. *"I am an Immortal Woman."* Focused, deep, rhythmical breathing will get you there instantly, and your relationship with your breath will take you instantly to "now" thinking, releasing your need for the past patterning to continue to reinforce its grip over you. The greatest gift you can give yourself is to silently breathe your mantra. To silently breathe means just that; stay in the stillness of the present moment with full awareness on the breath. The absolute power and joy you will receive is the living now, in the conscious moment. Really drink in the breath, as though you were a thirsty dog lapping at its dish of water; imagine yourself now as that dog relishing every cooling drop, your gleaming, pink tongue lapping every single molecule of this transforming liquid.

Studies have found that stress creates premature aging. To be old before you need to be is tragic. The results of living right in the present are as easy as a beauty treatment. Just think: every time I breathe with awareness and intone my mantra;*"I am becoming an Immortal Woman. I am one with everything; I am one with all that can be created now"*.

A memory of the past surfaces; you feel the sensation; you remember the anxiety, look at it like a cockroach that has run over your dinner plate during a meal you have lovingly prepared. What do you do? You attack the cockroach so it won't spoil your dinner. Thoughts of the past are cockroaches; eliminate them before they spoil your present. Examine the cockroach. Really look at it. This is your creation; you created that cockroach. The present cannot accommodate that past reality.

Immortality

Immortality is freedom, and for a woman to be in the present reality is her greatest fear. The present creates potential, and the Immortal Woman must always be on guard to create immortality in every second; every encounter can lead to a new awakening if you stay present. Make a promise to yourself that you will always stay in the present timeframe. See yourself on a movie screen as the character living in that moment, looking for connections in that moment, allowing life to unfold in that moment. Every single encounter is an opportunity to realize immortality; the feeling of it brings knowing that when you walk out of your door you can meet people, seize opportunities, and grow your dreams. Just think: every single encounter encourages your immortality because links with the past can surface to create new horizons in your reality. Immortality is living in the moment, creating opportunities in the moment, and being present in the moment through breath.

Breathing is the single most powerful action you can do consciously. It requires no thought, just intention. Say to yourself, *"I intend that my breath will be used consciously now to create immortality."* Breath is the most powerful action imaginable to create all that your higher truth needs as an Immortal Woman. My sequence of breathing first keeps me in the present; it acknowledges me as a living entity; it gives me life. Every second, my life ascends to immortal status when I develop conscious breath techniques.

The beauty of seeing yourself as immortal is very loving; you accept yourself totally when you believe in this way of living your truth, and "immortality" will be conferred on you. It's as though you will be stamped with this philosophy or way of being. All of life is a present moment happening; all of life offers the infinite possibility for your truth to manifest. Immortality says you don't see yourself as any particular age or status. For example, you don't define yourself by saying: I am a married woman. I am 40 years of age. Your culture defines you then. Of course you are married (or single or divorced), and you are chronologically a particular age. But your true definition of yourself doesn't have to carry this vibration. Rather, you can say: I am me living in this present state of reality right now. I accept all the joy and challenges of this present moment; this is immortality.

Our society does not allow a woman to experience this way of being—*of looking at herself as complete, an Immortal*—*because of the power it gives her.* Women give their power away too easily in our society, and the whole of society suffers for it. If a woman gives her power away to the system—the model she has found herself living in—she is, in fact, devolving that society. She must anchor herself to the Earth and know the Earth. Her Earth mother needs her protection and respect as well. For an Immortal Woman, life holds a knowledge that confers on her the wisdom of all the women of all the ages.

The Goddess

The Goddess of All Light is the Divine Mother, and through devotion and love of the Source of All Light, the Immortal Woman is born. She is born through purging the pain of living in a society that does not respect her truth, values, and ideals. The society of men is set against the creative principles behind all women. There is fear and disgust in giving women the power to nurture their beauty, and women react in turn by becoming bitter and distrustful of themselves. They see men's hatred toward them daily, and they, in turn, reject the very part of themselves that they need to grow. They then become emasculated; they fawn to male power, but they don't realize it is a trick to get their power away from them, to make them look like fools. How many women politicians have fallen for the cruel trick of the male hierarchy power only to find themselves "cast out" when they don't jump through the hoops of the power-crazy demons who run our societies.

The Immortal Woman gives these energies no power and will cease to allow herself to be a party to such destruction and desecration of all that is good and life-affirming in our world. Women on this path must recognize their power more; they must seek the truth within themselves and not allow themselves to be used by tricksters and fools. The status of immortality is one of "self" respect for all life and respect for the creative principles of the Great Goddess herself.

The Great Goddess, the Source of All Light, will always protect those who cherish her ideals. They are under divine protection. You need only to remember the power and life force of your own truth through breath work. Your immortality serves as a reminder of the continuity of all life. All of life continues regardless, and you need to feel strength and commitment to your chosen path, one of immortal service to The Goddess of All Light, the Great Mother, the Divine One. Your life is a testament to your desire for your truth to overshadow all obstacles on your path; just allow yourself space in your day to remind yourself of your belief in yourself and your truth. All of life is a testament to immortality, and all of life now allows you the feeling of the continuity of all living things. Your commitment to your truth brings life back into balance. There is no disharmony there, only truth, love, light, and perfect peace from the Source of All Light, The Great Goddess of All Light herself.

Passion

Passion for all life and living now creates this truth. Your passion for everything in life that is good and affirming brings this truth to you now. I am passionate about my truth, I am passionate about my ability to create and sustain all living things both inside and outside me. My life is a testament to the immortal cycle of existence, and I bring to my life my belief in completion and renewal in everything. I am a being for whom truth is made manifest, and I cherish myself enough to trust myself enough to be an Immortal Woman.

Peace

Just allow yourself peace at every moment. To bring yourself into a state of divine grace, you need just to evoke peace now. Keep saying, *"I am peace-filled,"* and imagine this peace permeating or radiating through your whole body. Peace gives a woman a sense of completeness in who she is. She is oneness when she is peace-filled; much can be achieved when the soul is allowed to feel safe and to radiate this sense of peace and love and joy of being a woman. The human road to immortality, which is present when the person radiates peace, is the only way; really there is no other way. Just find a loving surrender to the state of peacefulness now.

Peacefulness will give the body a chance to repair itself, as the old fears will not return. The fears of abandonment and loss will never return as the mantra of peace is evoked. The peace-filled woman radiates serenity and control and life force and hope to all around her; she expresses the divinity within her; her hormonal balance is restored, and there is no fear that she will have any emotional blowout as the glands are secreting hormones at the right balance. Breathe in the sense of peace and love for the soul now, and feed it your peace, and you will find immortality will be established. Your peace-filled life comes at no cost; you can evoke it by intoning, *"I am peace-filled. I am an Immortal Woman."* The immortality we bring to our life lies within us, and for humans, this is a very difficult concept to take on board, because we, as women, are so conditioned to believe and accept the words and actions of those who seek to control the flow of a woman's life force, drive, and power. When a woman hands over her authority to "the system" and allows those in control of the system to take charge, she is leaving herself very vulnerable to being depowered. It is just too easy to take something for a health change instead of trusting her own inner wisdom and truth. She must not allow her destiny to be shaped or shared with patriarchal organizations, which seek ultimately to destroy her life force and essence, and, in turn, will destroy themselves and our Earth.

Allow yourself space to understand you more. There should be time set aside just for you. A "you time" is very important to cultivate. You cannot grow if you are not in service to you. What do you need in your life right now to bring yourself to yourself? Any emotion and physical pain is an acknowledgment that you have neglected yourself. It is very painful for a human to give herself that very valuable moment to bring herself to find herself.

Treat yourself as though you are a child and you have to care for yourself. In other words, think, "I have a pain." Do not ignore this pain. Look at it. Where is it in your body or your emotions? Just find yourself, trusting yourself enough to be allowed to discover the source of your pain now.

Pleasure

The Immortal Woman loves pleasure; immortality is pleasure. When a woman is pleasure-filled, she radiates immortality. Pleasure is that sensation of experiencing life in the very moment. The very essence of pleasure is to experience life in its very special moment.

Think of a pleasure-filled person; she radiates a special magic. It is the magic of pleasure—the simple pleasures of living, loving, and experiencing life in its very special moments. Life is full of small moments of exquisite pleasure; you now need to experience life in all its pleasurable aspects. I give myself pleasure every day. Give yourself pleasure every day. Ask yourself, "What pleasurable experiences am I able to give myself this day?" List all the pleasures you have given yourself today: a good book to read in the sunshine or some powerful, deep rhythmic breathing. A human needs to pleasure herself; she just doesn't look or feel nourished without pleasure.

Your ability to pleasure yourself reinforces others desire to pleasure you as well. You only need to ask, "How can I receive pleasure unless I give pleasure to myself?" Do not buy out of compulsion or desire or greed. This is not pleasure; pleasure is the simple act of allowing yourself a deep moment of great love. It feeds the soul and doesn't cost money. Mantras are a very pleasurable way of nurturing the soul. I am now giving myself the pleasure of knowing I can please myself in small, loving ways. This way my soul is fed, and I can lovingly receive pleasure from others.

Uniqueness

Your gift to yourself is your belief in your own uniqueness. Your uniqueness creates the space of immortality; the knowledge that you can create in your life all that you want confers immortality on you. To be an Immortal is to totally surrender to the light. Your surrender to the light creates changes in the cellular structure to accommodate new opportunities for your life. Just imagine the power of allowing yourself the joy of total surrender to the power of light and it infusing your cellular membranes, re-coding your filaments within the microcosm of your very being; this is possible for all humans.

Multidimensionality

To be immortal is really to be able to accommodate multiple realities. We need to see ourselves as multipurpose beings; there are many levels and layers of reality, and we need to find within ourselves our consciousness changing to accommodate different levels of existence. Bring to your life now your belief in your power to just understand that you are the creator of your life and that this creation is now one of great magic, peace, and hope for yourself. Your immortality is the state of being in a space of knowing that you exist on many planes of existence all at once and that what you project or give out to others will enable them to be in a space of their truth. Just know and understand your power; all of the space of creation is yours when you create your immortality. You can be in many different realities all at once. All at once, you can be assisting those in other spaces over the globe and beyond. I am immortal. I want to feel my cellular structure and find new information about me and my reality. My reality is multidimensional. It incorporates all levels of existence.

Abundance

To bring the life force to you for immortality, you need to begin to allow spaces in your day to be in the presence of light. You need to breathe in the light and allow the light to really be infused in your whole cellular structure. Feel the tiny filaments of light encoding you totally. These tiny light filaments are codes of power and energy and will bring the abundant spirit to you. What is Abundance? Abundance creates life force and the desire for the very best life can offer you. Expect that life will offer you the very best you deserve by living wholly in the now; you are actively creating an enormous opportunity to radiate power and life force. There is a magical presence in living in the now. It is a life force made manifest, truth made manifest; and to allow yourself time to bask in the magical presence of your own uniqueness, is to live life as an Immortal.

Immortality is a state of mind, a belief that you can create this powerful transformative change of consciousness. The belief is the hardest thing for a human to achieve because cultures dictate that there are stages of growth and disrepair then finally death. There are cultures that do not hold this view, and there are people who live very long lives. Create what you need to create; ask yourself, "What do I need to fulfill on my journey now on the Earth plane?" Visualize a road; what do you—your higher self—want? Most people want peace, happiness, and contentment. To achieve this, they need to be able to release, and for a human, this is the most demanding task.

Release

Trying to teach a human to release is one of the most challenging paths one can take. Trying to teach yourself to release again requires dedication and a commitment to a higher truth. Your higher truth must always be evoked when you are trying to release. Release is the most profound undertaking a human is capable of.

To release requires total surrender to the unknown. The unknown has nothing, and this is why it is so hard to be human—we are wired for fear of the unknown. The minute a fear comes in, an initiation has begun, the soul has activated the challenge to the subconscious mind and ego, and all the unresolved karma then become activated. Analyzing these emotions, thoughts, and physical symptoms helps enormously. It is a question of charting the responses to the challenge.

I now wish to challenge the belief that I cannot be an Immortal Woman. Now note the fears and feel where they surface in the body: note the physical fear and which part of the body holds the fear or negative belief pattern. Check the body part; it may be the heart, throat, or any other organ or extremity. Memory at a cellular level is "stuck there." Note the accompanying emotions: Do you feel angry? Or is the inner criticizing parent saying, "Don't be ridiculous; everyone ages and dies?" Challenge the fear in whatever form it presents itself. This may take some time, so be patient and, above all, be kind to yourself during this process.

Breath

Note your responses; evoke The Goddess of All Light, the Source of the Divine Feminine; and self-nourishing breath is the most important function for immortality. Breath is a function of being in the present, so a breathing sequence should be established while examining the process of release. Breath creates space, and you can apply breathing techniques in all situations where fear is. Just releasing to the breath will create a profound peace and joy in the moment. This breath allows balance for the subconscious to process the new shift, the new experience of accepting the reality of being an Immortal Woman.

Just witnessing your breath, just witnessing "where" and "why" you hold your breath, creates a space of profound insight to your very being. Cellular reconstruction can take place where the breath work is engaged. Allowing yourself "engagement" in the breath is in itself a transforming process, one which will change your life forever.

Allowing the sacred essence of the breath to become part of your everyday reality creates a powerful electrical charge to your whole consciousness. This consciousness can be changed forever through the art of deep, rhythmic, regular breathing in varying sequences. A Pranayama teacher can help you with your breathing techniques if you find sequencing the breath difficult. However, the Immortal Woman—Divine Feminine breath sequence used is as follows: breathe in to the count of eight, hold for the same count of eight, and slowly, rhythmically release to the count of eight. This breathing sequence defies the process of degeneration in your organs and gives you instant vitality. The Immortal Woman must be a witness to herself because she carries the seed for her own evolution within her. Witnessing yourself requires full attention to your thoughts as they occur during the day. All thoughts must be on the process of surrender. Again the process of surrender must be one of allowing your divine essence to flower when all is chaos. When all is chaos around you, just surrender, because nothing can be achieved with chaotic situations. Chaos can come in many forms. Chaos is response to mass fear; your own fear is radiated out, and it brings fear to you.

Your need now is profound peace and knowledge that the quest for immortality lies in the present. The quest for immortality lies in your reaction to what is happening around you.

Challenge

Ask yourself, "What is happening around me right now? How is my life going? Who is in it that feeds my fear?" You must challenge your fear. Challenge your fear to know that you have the power within yourself to have exactly what you want. So if what you want is money, just be ruthless on yourself now to explore all issues around your "abundant" identity. See your abundance as not being related to your cash money, and you will tap into a reserve of power you just couldn't imagine. Just feel your power now to even "go there" around your responses to money (or lack of it). All around you is hostility, and you must recognize this hostility to your Divine Immortal truth. You are reacting to contraction and loss when you respond to such hostility; you are an Immortal. Say, *"I am immortal! I now encode immortality on me."* The fear of loss and failure robs you of your belief in your own immortality. Allow your belief in exactly who you are; be your own guide. Say, *"I am just so protected from the hostility of those around me."* See your body with a membrane all around it now, just letting light in and not letting hostility near you. You need to place this protective membrane all around you now, and continue to use it until you develop your own defense against hostility. What is hostility? Hostility is any negative projection toward you. Look at the media. The media projects hate; it offers no solutions, no understanding; it only offers fear to feed *your* fears.

Communications

Now look at communications. Look at your link with communications, and see who your "friends" serve by their communication. Really analyze *why* you draw people to you. You draw people to you to magnify your own fear; your *mind* constructs the need for friends to distance you from the power of your own truth and wholeness. Your ability to "create" friends is a distraction from your truth in your ability to become whole. You lose out by surrounding yourself with people who magnify your fear and chain you to a "mortal" life.

Truth

Life is a measure of your truth to yourself—it is a yardstick—and you must measure your truth to yourself daily. How have I fed my truth today? You must feed your truth by always acknowledging to yourself your small untruths. We give ourselves over to our untruths when we feel out of sorts with our hearts. There can be many moments during the day when we are asked to challenge our truth. You will know when you are being compromised. There is always that tiny niggling *anxiety* or *anger*; there is always a negative acting out over your psyche somewhere.

Just remember the last time this happened. *Yes*, you can remember. Now the moment is upon you. You must just stay calm and present and you *must not respond* as though you are under pressure. You need to gracefully give yourself time to consider the offer, proposition, or whatever doesn't feel right; responding right away only allows for your position to be compromised.

The challenge of immortality is always there because of the nature of being human. It is this awareness of your truth, which has been discussed earlier, so, you know it is always going to be challenged, because of the nature of being human. So just know exactly what the challenge is on a daily basis: What is my subconscious going to throw up at me today? How has it constructed itself to trip me up?

Being human is recognizing the inner demon and saying, "There is the subconscious playing up again; yes, I expect a performance." The subconscious will act on any previously held belief about you, through your culture, family belief patterns, social conditions, and whatever else it can use. So whenever there is a block in your thinking—a fear or a challenge—just challenge the subconscious mind to release this expectation of its behavior for your soul's journey. The soul has the blue print for the incarnation; so let the soul aspect really forward itself.

Resistance

Know there will always be a resistance. You must challenge this resistance, and you must allow yourself the knowledge that *you* are now in command of *your* life. The mind will always be there; just accept this as the challenge of being human. The mind is there to create diversion, division, and separation; your mind, then, attracts other minds of a similar frequency and need. When you feel your life is "out of control," you have allowed your mind to overload the circuitry in the brain, and it has become jammed or locked.

The higher mind cannot be accessed when your mind does this; you need to gently unplug or unhook its connection to your brain, so you will be freed from acting out its demands. Breath is very important, and it buys time for the higher mind to hook in again and create peace and balance in your outlook.

New Opportunities

Your opportunity now to create immortality will begin to have its rewards. For example, if you have decided to read this far and feel that you could find within yourself a "space" from your controlling mind to embrace the Immortal Woman totally, you will begin to sense a shift in your awareness of yourself in relation first to yourself.

Say, for example, you have begun to breathe consciously and intone, "I am an Immortal Woman," and you have done this for approximately one month, you will find that you are not so critical of yourself. You will begin to sense the subtle changes in your psyche, and you will find that there is beginning to be a new peace around you.

You will begin to feel more alive and more energetic; perhaps you will have made some minor dietary adjustments, an acknowledgment to yourself that you are really feeling just that little bit different.

These shifts are the re-coding of your cellular structure to encompass new frequencies. The new incoming messages are going straight to your cellular memory, which is making adjustments accommodating this new information. Every cell is beginning to respond to the belief that the body, mind, and spirit is immortal, that the diseased and degenerative state doesn't necessarily come with advancing years.

The focus now must be on allowing yourself to experience a new shift in awareness; the new shift in awareness will bring changes and challenges in relationships. All relationships are a challenge to stay centered and in your divine truth. When you start making shifts in your awareness such as this, there is now a need to understand that the tree is being shaken. You are being asked to release dead wood. Dead-end relationships must go; only those relationships that are evolving your soul's path, as an Immortal Woman, will survive. The shift will have a profound effect on who you are, and you may be accused of being "selfish;" moreover, you will probably accuse yourself. You will ask yourself, "Am I selfish for wanting more time alone? What is wrong with me?" You will be shifting your definition of yourself. Others will notice and sense your detachment and lack of need to hook into another person for security and protection.

Relationships

You will need to examine *why* you have relationships with people and whether they are fostering you and nurturing you at a soul level. Just really feel your need to be allowed to develop and nurture your relationship with yourself. It's such a shock for women to really acknowledge that they don't have to maintain family relationships (especially with in-laws). Women are conditioned to maintain relationships, and the Immortal Woman patterning changes this. Of course, the source of primary relationship is with the self, and your "self" always needs acknowledgment of *your* truth or identity. The self is such a powerful energy to focus on because without your identity firmly anchored in your present reality, you will become more soul damaged. A soul-damaged woman cannot be immortal because immortality requires total integration of the self.

Integration

To integrate the self is one of the biggest challenges in being human, and the acknowledgment that you need integrating is a very important step in attaining immortality. Ask yourself daily, "What parts of my energetic structure need integrating now?" Breathe deeply, powerfully into your "energy field," and see it fill with light; at this instant, visualize where there are "holes" or dark spots on your aura or energy field. Visualize yourself like a Christmas tree with many lights of different colors. Ask yourself, "Where in my energetic field is there a lack of light?" Now find within yourself space to find where those dark spots can be fixed. A picture will be shown to you; you may begin to see a light appearing as if by magic; you may even see the very color. Now let yourself bring a glowing light to this area. It is very important that you just see this magical, glowing light now; you are on your way to being integrated. Try this whenever you are "in the dark" over something or someone. Just allow yourself to find the colored lights so your Christmas tree can be lit up, so you can be integrated and made whole.

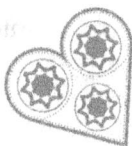

Chapter 7

SELF RESPECT

Allowing yourself the space to love yourself is an awesome challenge. It's just so hard for a human to do this. The subconscious will resist every attempt to love the self enough to give space to just acknowledge to yourself that you need to be with just "you." Women find the prospect of just being alone with themselves a special challenge because of their responsibilities, which, in the past generation, have become more burdensome. It is just so easy to give the soul space to grow. As a baby needs sleep and much of it, the soul cannot grow in spaces of responsibility; the soul can only flower by attention, and the space must be quiet, reflective, and introspective.

At the minimum, give the soul the first hour of the day to begin your journey. This can be a slow, meditative awareness, a chance to process nighttime dreaming, exercise, or music. The idea is to introduce yourself to yourself; every day you need to introduce yourself to yourself, because you have grown in 24 hours.

Allow for evening space to reflect on the day, and release the day slowly—just wind it backward. Take it apart, segment it, and let it go. Bless it; dismiss it, and *let it go*, just like a teacher dismisses a class at the end of a lesson. Try this blessing and dismissing at the end of every event that has been important—a project, a person, a job, a house, a marriage, a child—and always forgive yourself for any karma and ask to have this karma released from all lifetimes. You then will not re-create a desire for the same situation, person, or event again.

Saying you are an Immortal Woman is really a challenge to all women to define themselves in relation to their currently held beliefs about aging. Saying *"I am an Immortal Woman; I believe in immortality"* gives you a chance to define yourself. Don't just fit into your society's definition of aging and decay; being immortal means working on your belief system about aging and death. Every day you can say to yourself, *"I am an Immortal Woman"* to recode your cellular memory with new frequencies. These frequencies are being restrung with information that is new to the cellular structure. Breathe in every day. Say, *"I am an Immortal Woman; immortality is mine."* Every cell then has a chance to renew itself, so it won't respond to its currently held belief about aging and decay. The possibility of aging is then foreign, and you will not perceive yourself as being old or go anywhere near that definition of yourself. Your life is now one of creativity, and life and your cells will respond to your creative impulses. All of life is a source of creation, and the creation is being infused in your cellular structure the moment you see yourself as an Immortal Woman.

Self-Adoration

To be in awe of yourself is the most awesome thing a woman can do—she sees herself as all that creation is capable of. Her creation is all about herself. Herself is her worshipful behavior of herself. Every time you look at yourself in love and adoration is the moment of magical creation. You are capable of this self-adoration; your belief in the knowledge of self is the sum total of *you!*

Essence of Self

Allowing the essence of yourself to be explored by yourself is the secret of your immortality. Just see your essence — the distilled part of yourself — as being a spark of your divine truth. You need to explore the spark and really find a space within yourself where the essence of you is explored. Your essence is the secret spark; essence is essentially the real you — the real you that needs to be acknowledged. The real you lies hidden, buried away, and unable to be explored at all.

We are afraid to explore and expose our essence. Our essence is our secret, and secrets are not likely to be developed when we live in a space of restriction. Bring to yourself your essence; ask yourself, "How can I know my essence, my essential self? What is it?" Then tell yourself, "My essence is the secret of my immortality, so what do I need to take with me in its intense form? I love myself enough now to bring this essence — this secret part of me — forward."

Devote some time to really acknowledging to yourself that you have an essence self that needs acknowledging for your immortality. Say to yourself, *"My essence self is now being explored."*

So let's explore our essence selves now. Tell yourself, *"I now need to know that I can explore my essence self in meditation."* Ask the Goddess to show you your essence self and bring it forward now. Begin your breathing sequence of eight in; pause eight; eight out. Then say, "I now allow my essence self to be revealed to me to attain immortality."

The beauty of your own life is the Goddess within you; your life must be a model of the aspiration of your totality, and to feel the loving presence of the Goddess within you, every moment, is your testament to who you really are. Your statement is your truth to the world. Every compliment must be acknowledged with an acknowledgment of the Goddess within you. Thank silently all the forces assisting you in attaining immortality. Your moment-by-moment experiences will allow the essence of who you are to come through.

CARMEL GLENANE

Just feel the sense of profound peace, which comes with knowing *you* are a child of the Goddess; the belief in immortality is just that. You must really and truly believe that *you* really *deserve* all that you want for your journey of truth and awareness. The journey of truth gives you strength to cope with the trials and battles ahead. The knowledge of your uniqueness and completeness is yours now to really look into.

I love myself enough to bring the essence of the Immortal Woman to me for every "act" of creation. The journey now is one of trust and true surrender to the forces that serve you at this time. Life offers many rewards for adherence to this truth: the need to stop and not push so hard to create your personal statement of immortality. Ask yourself, "What can I do this day to create a space of immortality in me now? Where is my truth in my life right now to create immortality? Say, "I ask for the truth of immortality to be revealed to me now. I am in a state to receive immortality now." Be in loving surrender to yourself now and go about your day with awareness of your essence through immortality.

Allowing yourself just "to flow" brings the state of immortality to you. The flow of life begins with acknowledging yourself as having space to ponder and reflect on life. Just allow reflection and the knowledge that you deserve to just float or flow with your life in its "essence." Ask yourself, "What part of my essence am I not flowing with at this moment?" Tell yourself, "*I am free to flow and float. My essence is the spark of creation that has no beginning or end, it just has an essence.*" The essence is the "you."

The *you* in your uniqueness brings yourself into a state of reflection. Stop and ask, "Why am I not able to float with my essence now? My essence needs freedom. The true essence of me is my ability to just feel what I want for my present state of awareness." You need to float with your essence self. It is like watching on the water a seed; the essence is the seed of your creative self—the self that is limitless and boundless and free. This space of loving surrender and peace within yourself is a good space to observe "you" from—a space from which you can just float and let go of all you have created from your mind and its prison, allowing yourself to ponder the space of creation, so you can just float in your true essential form.

Trust

Life brings many opportunities to test yourself in relation to your humanness. The first challenge is the trust process. Trusting in you is one of the biggest challenges of being human and, therefore, your immortality. You are the truest source of power and respect for that power, which creates in you the abandonment to the trivial and the mundane. Never allow yourself the knowledge that you are merely mundane and, therefore, not worthy of the very best your circumstances offer. The true test of your immortality as a divine incarnation of the Great Goddess is the knowledge that you can trust and believe in yourself enough to listen to your truest and most unique source of wisdom: *you*.

You confer immortality upon yourself in every life situation where you are truly trusting in your own self — that self of love. Love is the most neglected word in the human language; yet it is the most abused word also. Why? Because love is the seed of immortality, and it is the greatest challenge to your personal greatness and self worth. The energy extended on loving yourself is the best energy you can spend on anything. Selfishness is just the opposite of self love; selfishness is the inner child needing love. Selfish behavior isn't an act of self love but a betrayal of that love. Your immortal status is not a selfish one, but one of the "self." The self is the most wonderful creation imaginable, and you will feel the pull of "the self" to create the self-*art* of love, which is immortality itself.

The beauty and joy of trusting cannot be underestimated. The joy of trust is like a child's simple response to life. We, as humans, must develop that trust in all that we do and what we expect from our life. It is just so simple to say, "I trust you." When you do, you are bringing back a soul part of yourself that has been chipped, cracked, lost, split off, or just gone. Every act of trust is a trust in your own wholeness and love for yourself.

Continue always to monitor your trust of yourself when being presented with a challenge. Tell yourself, *"I trust my inner guidance to trust this situation, person, or whatever it is that I am facing."* Always get the permission of the inner self — the true self. Trusting anyone is just like trusting the true self to reveal herself. Just let her come out; she is a frightened child; show her that you can look after her and wait for her response. When you trust yourself, nothing can harm you. So don't act impulsively in a situation that is new and challenging. Just give yourself time to develop a relationship with yourself. Then and only then can nothing go wrong.

Your responsibility is to always nourish that source of trust that lies deep within you. You owe it to yourself, your source of power, through the trust process. The process of trust brings you into alignment with the forces of light to assist you. Your trust channels light beings to you to further aid the trust process. List all those you have been able to trust and see how the list grows. Do not concern yourself about who you haven't or couldn't trust; forget those people. This is not important. The trust process is the *truest* process of living. You cannot live without trust, and you must continue to trust your own truth. An Immortal Woman cannot live without the truth, and the truth cannot be revealed without trusting.

Imagine a baby not trusting its parent or caretaker to give it all it needs for security and nourishment. There is no analysis of thinking *why* will she not come to feed me or otherwise care for me, the brain hasn't yet developed the critical faculty that questions. To trust, you must get away from the incessant questioning and the obsessive "why." This destroys the trust mechanism. To destroy trust in yourself is to deny yourself the true gift of being human and of being an Immortal Woman. If you trust someone not to hurt you, you will send out a vibration of love to that person, not your pre-loved subconscious fear, to which the other person would respond by not flowing with you and not allowing you to trust in him or her.

When you trust, you don't have the issue of emotion with another person; you just acknowledge that person as a trustworthy human being. If you are betrayed, you are not really betrayed at all.

That person has betrayed some aspect of him or herself that wasn't healed, and it is his or her problem not yours. You are never hurt when you trust; only the person who betrays you is hurt by not honoring something in him or herself that was out of alignment with his or her soul's path.

Live Now

Life, when lived in accordance with the philosophy embracing the Immortal Woman, is full of absolute joy and release. There is no past, absolutely whatsoever; all you have is this second, and that second is all you have. Tell yourself, "I have only this second now with this person or concept." Ask yourself how you would feel now, at this second, when you are able to tune in to that person or concept. All potential in the relationship lies now, not in the past. The past just doesn't exist; just imagine a relationship you have been struggling to maintain because of loyalty or abandonment issues, either your own or the other person's. Now just imagine *now*: I am meeting you for the very first time, I haven't seen you before, and I know absolutely nothing about your background and history. I look at you now. We begin contact; do I want to be with you *now*? Breathe deeply three times; really breathe from the belly. Now breathe from your heart. Now breathe from your throat area. Breathe in and out. Then stop and ask yourself to co-create here in your divine truth, and in your divine truth now, find the answer — yes or no; it's that simple.

Either I wish to co-create with you or I don't. Now, you can do this for anything. Look at your home; you have been at the same address for some time. Try the method. What happens is that you'll have an instant answer — the way to your truth — and the way to immortality lies in this surrender. "What am I surrendering?" you might ask. You are surrendering illusion, fear, pain, abandonment, false loyalty, and loss of love for the self. So where is the problem? The problem is in finding the answer where the problem is. The answer is too demanding to deal with, so don't deal with the answer all at once. Get help. Help is on its way; because you have embraced the philosophy of immortality, the immortals, the shining ones are now ready to assist. They are there for this very purpose. When you evoke them, you evolve them, yourself, and all of creation. Everyone takes a step up the evolution ladder.

Mask of Life

To align yourself to receive light daily creates in you the ladder toward immortality. Immortality implies ascension; ascension is (merely) a tool to create the immortal you now; to ascend is to create the consciousness of limitlessness. Limitlessness creates immortality, so to be an Immortal Woman is to take off the mask of fear; it is only fear, which limits the spirit's ability to be immortal. Visualize your mask. You wear your mask every day; you put it on to cope with the fear of being human. So breathe the mask away; see it dissolving before your very eyes. When the mask of fear has gone, you are free to be an Immortal. Before you begin your day, you need to just breathe the mask away. Say, "I now lovingly embrace immortality. By breathing away the mask of fear I wear every single day, just for this day, I will have no fear. Just for this moment, I will feel free to be fully released." You don't need a lifetime of fear around you. Just decide each day that the mask of fear will not create in you the illusion you see around you, because when you are fearful, you are wearing a mask. Now who wants to see a mask instead of a face? You need to show your face to the world, not a mask. It's an insult to greet another human with a mask. Like a fancy dress party, the masks all come off at the end of the party. At some time, you really have to reveal yourself, because people want to guess who the real person is.

The real person is the immortal you. The immortal you knows no fear, so when the mask is off, you are on. You are open to receive: no one wants to give to anyone who will not open his or her face to the world; no one wants to give a gift to a masked person, when you link to a person. Look behind the mask to do this. Look at the eyes; the eyes are the contact point; just think of a mask. All you can see are the eyes, and you look to the eyes as a point of contact. The eyes are the only opening to the soul. A person's eyes reveal much; they will tell you if he or she wants to take off the mask.

When people close their eyes, they do not want to reveal their soul; they wear their mask. True intimacy is in connecting at the eyes. Do not be afraid to look directly into a person's eyes if you want to see the soul.

The state of the soul is revealed in the eyes. Yes to eyes.

What can you tell about yourself through your eyes? What level of immortality are you at right now? Your eyes will reveal your level of immortality. How do your eyes do this? Can you tell whether a person is evolving consciously?

When a person can gaze lovingly at you, as opposed to longingly (which in itself indicates a need or frustrated desire), you know they can be in a space of consciously understanding themselves. Life offers a multitude of experiences to create a space of immortality, just by observance.

Observation

Do you find a clue to your own creativity and immortality? A creative person observes. Observe all around you. Ask yourself: When was the last time you really observed the stars at night and connected to their form and majesty? When did you really look at the daytime sky, observing the cloud formations and patterns they make? When was the last time you really observed nature, a tree, a flower, a shrub, or a bird—really looked at it and observed it? Observing is not just looking. Observing is engaging the senses, really engaging and pulling apart their form. It's time to analyze their form, and test the form against your life.

The energetic imprint of the cosmos is also ours as well. Its doctrine of signatures can give you a healing, just by observance; simple observance of the natural world creates healing. You can heal yourself and create change at the cellular level of your energetic makeup by observance of all around you. Ask yourself, "When was the last time I really observed?" Instead of looking, now observe;

observance makes for healing of oneself. Now observe someone you love; what do you note about them; what is "it" that you find? Really observe; when you do you will discover a new creativity and depth to your life, which in turn will bring depth, color, and passion to everything you do.

Say and *create*...Creativity equals immortality; Observe and create and bring immortality to you now! Observe yourself; really observe without criticism; really get to know yourself.

Do mirror work ...begin to observe your face, hands, body, feet, and legs. Don't judge; don't criticize; just observe your form. This exercise is merely an opportunity for self discovery. Discovering the self, *there is no judgment of the result.*

Choices

The challenge is upon you now to sit up and really decide what you want for your life in human form; it is the creation of your humanness and the acknowledgment of it that brings your personal quest for the status of immortality. You created your human form; it was you who created yourself and chose to be human. You had this choice; you made it in a space of completeness, but you wanted more. Human life teaches us certain things. The soul asked for the challenge, yet it recognizes that the human condition is one fraught with *perils and dangerous undercurrents.* It is like a cork floating in the ocean of humanity, and it needs love, protection, and support on its journey to grow and learn in human form.

Receive Love

We had all we need as babies—we had love (hopefully), and we had protection. If we did not get those things as babies, we left ourselves very vulnerable to the dangerous undercurrents. If some of us did not survive those undercurrents, we were maimed psychologically. You cannot survive without it; so you must recognize that to be human and, therefore, to be Immortal, is to receive love.

How do you receive love? How does a baby receive love? A baby receives love by being herself; she is just that...herself. She draws love to her by being herself. A baby does not have falseness; she doesn't have an agenda; she just trusts that someone loves her.

Begin the process of love in you now by acknowledging your worth to yourself *now*. How can you expect others to acknowledge this worth if you don't do it yourself? All of life is a moment-by-moment journey to love the self. There is a challenge every second. All of life is a challenge to bring the loving energy of the "self" into alignment with the Goddess, who creates the life force in the human spirit. Just practice daily the process of accepting life as a challenge to love the self. All of life is a challenge to love the self now, and all of life brings the moment of being human, transcending it to being *superhuman*.

Regeneration

The idea of regeneration and renewal is at the core of immortality, so when you regenerate yourself or renew yourself, you are creating immortality in you.

When do I regenerate myself? Think about when your personal rejuvenation takes place; you will regenerate yourself by deep contemplation alone. Your process of regeneration can take place in observing nature and bringing the forces of the natural world into your very being. Think of the season of regeneration: springtime. The feeling of spring is renewal. *A sense of new life is the wellspring of regeneration.* Breathe in the season of regeneration; breathe in springtime. The season of the abundant, natural world is at the core of a renewal in you, and this sense of renewal and regeneration will bring with it your sense of well-being. *A sense of being a positive, regenerating energy enfolds you now.*

Affirmation

I am unfolding in my ability to regenerate myself by bringing in the positive health-giving benefits of springtime, the season of regeneration. Immortality is new life. I am bringing in new life to me now! What new life can I foster in me now? What newness is coming to me? I create new living things in me now! I am alive to creating newness in my life! What can I create today?

The process of creation is newness, freshness, and commitment to making something unique. Say, *"I am a unique creation myself. I am my own unique creation."* Create newness in yourself. Ask yourself, "Now what new way can I do this task? Complete this? Make this?"

Renewal and regeneration create commitment, and the immortality of the "self" is now allowing that regeneration and renewal to be a daily part of you. The process of observing all around you has no parallel reality; there is *no other way.* Just allow the spirit to merge into your totality through your breath, for your breath creates a "space" for renewal to take place. The breath is the sacred fire of creation. The breath is the sacred self-made perfect. It is just so easy and so amazing that a human's conscious breathing can create consciousness instantly.

Sacred Breath

The entire self-help industry could be transformed if each person who wanted to help him or herself could just acknowledge the sacred alchemy of the breath. Immortality is just a creation with raising the consciousness, like a television antenna. Just let the breath guide you and allow it to draw sacred pranic energy to you now for your life in the creation of oneness with all of life.

Be in celebration of the most life-affirming power of the sacred breath. The sacred breath is the might and light of the spirit and the soul, and it must be acknowledged at all times.

Be in wonder and creation at the wonder of the breath in everything that you do, say, and feel: the breath just makes for the status of immortality, and the immortality is the creation of oneness and newness in all of life. All of life is the sacredness of spirit made matter. All of life creates sacred oneness now.

I am safe, I am alive, and I am immortal, because I am consciously breathing!

Self Realization

To begin the journey of "self" realization is the quest to re-define yourself in relation to all around you. Re-defining yourself to all around you confers to you the power over the useless emotion that creates more and more karma. The emotional attachment creates the karma, and it is the karma that *you* are reacting to in every life situation. Every experience is a challenge to balance karma and restore a state of inner harmony; inner harmony is the state of oneness of bliss with all there is.

Balancing karma is the challenge of being in a human body, and you need to challenge yourself now to bring to your life all that there is to grow your truth. Your truth is your belief in the status of immortality and, therefore, eternal life in you and through you. You need now to allow your divinity and oneness to be realized in all that you do, say, and feel.

The essence of immortality is just that—the essence and oneness of your totality. All of life is a celebration of oneness, and you now need to find the expression in the totality of who you are through release. Releasing all emotional attachments creates immortality. I am the essence and divinity of the Immortal Woman. I am the goddess who exists in all of life; all of life is one with me, and I am creating immortality in myself through observance of my total being. In oneness with myself, I am immortal to the level of my consciousness.

Freedom

The Immortal Woman is the essence of freedom; you cannot be immortal, in other words, free of disease and death, if you fear what disease and death can do to you. You may fear a disease; for example, in our society, cancer is the dreaded disease. Yet by fearing it, it in some way, becomes a magnet to you, magnetizing itself to you. So really, to be free is to be non-identifying with a particular condition: in other words, my heart isn't good or our family has a history of heart disease. Your need not to identify with such conditions creates in you freedom. Say to yourself, "I am free from identification with such a disease; I am free not to feel the connection with such pain." The pain of being human confers on the spirit the need to identify with such pain, fear, and abandonment. To be free is to just not identify with such pain. I have no identification with pain, fear, or loss associated with disease; it is not part of my reality or my creation. I create in love.

Death

What about death, just what is death? Death is merely an illusion created by the mortal man. The mortal human fears death because he or she wants identification with the source of being human, in other words, being in a body. To be fearful of death is to be a slave to the body and all its identification with pleasure, desire, and other physical needs. When the body craves something, for example sweet food or sex, remember that you are in the process of allowing yourself the responsibility not to identify with it; in other words, you can have it, but to identify with it diminishes you and creates the chain of mortality. I can eat a chocolate or a box of chocolates, but it is the identification and the need that is the "mortal" in you wanting to stay "human."

Desires

Your cravings and desires as a human always need to be addressed: This is why ritual is so important. The ritual of doing something creates space for you—the ritual of knowing you are able to observe a routine, contemplation. It is a good time in your time to "routine" it, as conscious intent with a routine creates a sacred ritual. A sacred ritual creates a powerful space in your life *now* to bring to you *all* that you want for *your* life. Your life now is one of ritual, to surrender all desire that does not further your immortality. The Immortal Woman needs order and ritual to create space so that fear and desire can go.

All fear and desire are only the need for the mortal woman to stay in control. The mortal woman always wants stimulation, always wants desire satisfied. It is important not to satisfy desire with stimulants, as they only require the soul to begin the journey all over again. The chain of immortality is just that—a chain. Say, *"I am constantly releasing to the link to immortality in everything that I do. I feel the chain of immortality; it links me to cosmos, the heavens, to all there is."* All of life is an event, and your relationship with your immortality brings to you a sense of knowing you are in the right place and time to bring all you wish for your growth right now.

Source of Power

Allowing the feeling of energetic connection to your source of power is tapping into a great reservoir of strength in you to create your "immortal" status. Having spiritual direction or focus is very important in maintaining your link with your core beliefs about immortality. The guidance of an ascended being who is there is very important in your quest for immortality. The divine being is *outside* you and guides you, protects you, and supports your newly emerging identity to spiritual awareness. All of life is a celebration of this connection with a spiritual tradition: It does not matter what tradition it is, or *how* you choose to *connect* with this being; just allow yourself to grow this spiritual connection. All great faiths have had this spiritual being—a god being or goddess being. Feeling the knowing of this loving connection just feeds the "divine" energy of knowing and trusting in this energy to guide you at every moment. The loving trust of an entity helps you; now just allow yourself to be infused with this loving, embracing energy of love for "light," as the being must always have a light source.

The enormity of knowing there is a universe of love out there supporting you aids immeasurably in your attainment of immortality. Tell yourself, "*I am now surrendering and trusting this divine source.*" Bringing this source of universal energy to you will create a very valuable link to your immortality.

Linking with the Source of All Light by evoking oneness and commitment to you creates in you your special relationship with yourself. You need to flow on now, just allowing a sense of peace to enfold you; every living moment, every second is an opportunity to give peace a chance. Give peace a chance to live in you. Be aware of the karma of daily challenging yourself to *really open up and embrace you – to always give something to yourself first.*

Giving

What you need to ask is this: *Did I give to myself today?* I cannot begin to connect to my sense of immortality today. I am an Immortal Woman. I give myself a chance to be immortal through my love of self creation: what part of my "self" am I creating *now*? I am creating my self-worth; I am really connecting to my sense of self-worth through the connection with the goodness of my life.

Much of my life is "good" and the "good" parts must be fostered and developed. The sense of goodness now is a belief. In your true worth and your sense of self-worth, make your "self" a worthy self, a self that is connecting to the goodness in you. Your own belief in your goodness creates a climate that allows the self to grow. If you can't believe in your own innate goodness, you will need to allow yourself space to connect to your goodness. List what is "good" about you now!

What is "good" about me now:
-
-
-

All Love

The energy of love around you creates in you your sense of immortality. All of love is the essence of oneness with the universe. Surround yourself with love. List all those who have ever loved you and remember those people. Love needs to be measured; being in a space of surrender to love is the merging of true oneness of spirit and essence with all of life. Just co-create with the essence of love in its totality. Through this essence, you will become attuned to magical happenings in your life, and the goodness from others to you will only magnify your goodness to others, allowing the spirit of love to infuse your every waking moment. It is the spirit and essence of a life that embraces immortality.

Your total commitment to bringing to your life a way of "being" creates in you your own immortality. Just allowing the way of being to infuse your every moment is a creation to bring the energy and essence of immortality to you. Bring to your life now the joy and magic of immortality by just recognizing that you and your creative essence are co-joining in oneness of spirit.

Your life force essence and oneness is the journey you need to make now in oneness for all of life. All of life is a creation of oneness, wholeness, and spirit; so it is with immortality. Being the state of shared oneness with all of life to you now through the goodness in your heart, embracing all of life is a component of all of the essence of oneness.

Life is a rich river, and the energy of allowing this river of oneness to be part of you now is the challenge to embracing this oneness and aliveness. See yourself in a deep, flowing river; all of life is swimming along; there are big fish and other underwater life, and all are floating along. Immortality implies that the mouth of the river flows into the sea and that the sea is a large, open body of endless water. Rivers are our veins, our life force; they must be kept open and flowing all the time — forever open to opportunities to grow, looking for snags or dangerous currents. There are dangers in the river, and the danger is in being oblivious to the currents and energies that will manifest at this time. Just knowing and living in awareness at every moment — just living with awareness moment-by-moment, hour-by-hour, day-by-day — is immortality. Know there is the mouth of the river and that the sea is open and endless — a sea of open endlessness, opening up to infinity. Visualize yourself on your journey now. Which stage are you on now in your understanding of who you are? Your understanding brings to you now your desire to fully engage in the process of life, fully and wholly engaged in life's rich process. See yourself floating, observing, partaking, and avoiding that which diverts you from your true course, the course of the endless sea of immortality.

Creation

When we create in love, we create in ourselves the Immortal Woman. The Immortal Woman is creating in love at every moment. She is engaged in this flowing. In the river, I create in oneness, and the great river of life is in oneness with me now. I am a creation of the great river herself, so I lovingly allow myself to flow with her and merge with her now. When you are open to flowing, when you go with the flow, you are creating the Immortal Woman in you, and your flowing is just that — a flowing. I flow with life. I am an Immortal Woman when I flow with life because this is the creation. Being stagnant, relationships, people, and ideas all must be constantly challenged. The constant challenge is to flow with life, and the flowing creates in you the merging into oneness, the merging to the Immortal Woman. She is immortal when she creates: so all of life is just that — a creation.

I create life in me now by participating in life, I am creating in oneness, and all of life is oneness. I am now a creative being flowing in oneness with all of life. The goodness of life is just creating, flowing, and partaking in the daily challenge of being woman. I can create in life by observation, observation of myself. I observe my essence and oneness. I am creative; I create when I lovingly acknowledge what I have. I am creative when I allow myself the opportunity to give myself the very best I can afford. I can afford the very best, because I co-create with life herself.

The Immortal Woman is the greatest and most exciting creation. Just acknowledging the Earth is an act of creation.

Silent Space

The Immortal Woman embraces silence; she always needs a silent space within her life so she can grow new aspects of herself and let them come through. She needs to allow herself the space for this to take place, and finding the inner space is very important for her day. Allowing yourself time to create that inner space means you need to be consciously aware of your time and to monitor what you are giving out to others; this will give you what you need for your inner life of silence, stripped bare of all pretensions and illusions. The joy of sharing yourself with no one but yourself is a call from the soul to recognize it, and because you are seeking immortality, you will need to give yourself this time to develop the sense of spirit within.

Your silent space allows for remembrance of things past, and you will feel a sense of recall and remembrance of long-forgotten truths about yourself. Your silent space embraces this wholeness; this remembering and this shared life you have with yourself will bring a rich reward to you. Going within and feeding your soul is a goldmine; you can find that your life will give you all you want: allow yourself time and space in your day to embrace total silence and allow the immortality of your soul to soar and flow.

I allow myself to embrace immortality by being in a space of silence, with no restrictions and absolute freedom for all of Life.

While you allow yourself your silent space, you must be active within this silence. This silence is not passive or sleepy. It is active and watchful. The silence must be watched. Observe your silence carefully. What intrudes upon this silence? Look carefully at this silent space. Allow the thoughts to just float and drift. They may land like a butterfly, but only for a moment. Just observe them and give them no energy whatsoever. This will create a greater volume of space of pure essential light and peace. You can feel that you can really trust and surrender in this light space.

This space creates the real you to emerge; the real you emerges now, and she is going to be the you that you truly know you are. You are now creating in your life a space for that truth to emerge, and when it does, you will be in surrender and love to all there is. Know now that you are able to lovingly support yourself in that truth, that it merges into you when you are released from your silent space and you are whole to find the spark of immortality to bring to your daily life your truth, love, and wholeness. Begin now to silently witness your truth; begin now to be in a space of shared truth and love for yourself so you can grow your immortality. Tell yourself, *"My immortality and truth are for me when I witness myself in loving silence."*

Inner Exploration

The belief in your own immortality creates in you the shift so you can really open up to the aspects of yourself that need to be explored and exposed. Both exploration and exposure create energy, an essence of truth for your immortal status. Finding a space within yourself is very important for this precious alchemy to take place. I just allow the space of inner exploration to bring to me now my journey of trust and shared truth. I begin the journey right now because I allow myself to really surrender to myself. Only myself can make this journey; only myself can find the moment of surrender and trust in the overall flow of life. Just allowing trust and surrender to be part of your new energetic makeup will create much love, peace, and hope in your life right now.

The feeling of just letting go and allowing yourself total belief in your own ability to really create the beauty of an Immortal life is worth every ounce of effort expended to be in this truth.

I now allow my truth to really be explored and developed through knowing that I am an Immortal Woman. The sense of my being an Immortal brings to me great hope and love for my future, my energy, and my life force. Self love is directed to this truth, and I am now in loving surrender to all of life. Just allowing yourself *now* to be in a space of truth and surrender to the love of immortality is your truth.

Loving yourself is always a challenge in your developing relationship with yourself; it is very important to really just try and imagine the energy and effort required to challenge so many deeply held patterns of belief about yourself. You need to really allow yourself the opportunity to feel this energetic shift of gears for your relationship with yourself. Just say to yourself that you understand and trust that you would be able to make this energetic shift in consciousness because *you* know how important it is to see yourself this way.

Struggle

You must try and understand the benefits of becoming an Immortal Woman; feel the benefits in your energetic makeup. You may ask, "What will happen to me if I decide to embrace the new way of being?" First, you will stop struggling; you struggle as a human being when you limit your thinking to the current way your society determines how you should behave or think about immortality. You need to develop awareness; you also need to learn not to necessarily follow rules just because they've been made. They are made to control. You don't have to break the law, but you need to *bend* the law. The only law you have to obey is your own inner truth. All other laws only serve to limit the human imagination; so gentle challenging of yourself in relation to how society defines "you" is important.

The Immortal Woman is a statement and a belief in your relationship to yourself; your relationship to yourself must be above all considerations, even loved ones. However, there is a responsibility to care, but the "care" must be balanced because you are serving your truth; your master can only be yourself. Your master is the only yardstick for growth, and that master is yourself; no other will do. The status of immortality will not hold it because you have allowed yourself space to grow that "master" you. The testament of your life is how you lived your truth, and the only judge of that is you, so it is essential to breathe that truth into yourself at every opportunity. Tell yourself, "I am *a belief in my truth*, and I acknowledge the truth in me and others. I am a statement of truth, and I acknowledge that I can create all I need for my life of immortality." So in the end, we judge ourselves. We know whether we hang or can be saved. The judges of the underworld are ourselves; we know whether we deserve to cross over the threshold to our immortal status. Every day, judge your actions; then the great god of the underworld Osiris, whose decision is supported by Isis and Nephtys, representatives of the divine feminine, will not need to judge you at all.

Birth

Birth is the journey of creation; birth is a new beginning. As to birthing the Immortal Woman, this is an act of seeing yourself as something new, something different. So why not just release and bury the old you, the you that is tired and not creative? Immortality brings an endless source of creativity because the universe is just that—creative. The creativity of your spirit now brings a new status to you: I am birthing my immortality now. There is a new creativity in my status. How do I identify myself now? How do I share myself with others? I do these things by being my creative self. Your creative self is your immortal self; your creative and mortal self can combine to create an Immortal self. You need the creative expression to create the immortal self out of the mortal self; the immortal self is then an extension of the universe of creativity. The universe of creativity is now able to be accessed through your immortal self. There is an extension, definition, and recognition of exactly "who" you are: the immortal and the creative, the extension of your mortal self, become one. Immortality is in tapping into that self; that self is part of the universe web of creativity. When you *see* yourself as immortal, you are recreating the web, the matrix of yourself in a universal form. You are not separate from the universal web of living energies; you are actually part of the living flow of energies.

The energy of creation is in you; you are actually allowing the flow to become one with you. *You and the flow of life are one.* The web of immortality is "in" you.

Just being in the presence of yourself is a creative act: The act of creation is in knowing that you and yourself are immortal. The act of creation is recognizing that you are capable of being a universal human being. Being a universal human being is the action and intent of being human; recognizing your humanness is very important, and going *beyond* it is the beginning of developing your immortality.

Humanness

Yes I am human: I am tired; I am sad; I am alone; I am fearful. Good, this is being human: it is now, in catching the moment and not submitting to this humanness, that is the beginning of something new. The new understanding of yourself now creates your humanness. Your humanness now goes beyond the normal range of submitting to human emotions. Just learn now to slowly, deeply, and rhythmically breathe, to slowly connect to the immortality of yourself, your blue print for immortality, your matrix. Your breath connects you to this source, so the need to fall into the fear, the emotion, now creates in you the desire to go beyond or transcend mortality to immortality. So the sacred breath confers upon you its threshold to transcend the limitations of being human, to begin the journey into oneness. Beauty, peace, and love, the journey of being human, is just that—the beauty of knowing you, the beauty of knowing you don't have to submit to being human.

To embrace yourself in love will establish the link to embracing your immortality. Just embracing yourself, enclosing or encasing yourself, in your own essence gives you the desire to acknowledge yourself. Acknowledging the self—the self, owning that self—just recognizing that self is you, creates in you power. Your power is to now just bring to yourself all that you, the living you, require. An acknowledgment that the self is worthy, that the self is whole, brings now your shaping of yourself.

Identity

Your identity must be firmly anchored in the self; the identity of the self is one of shared responsibility between the self and the emerging Immortal Woman. Sharing the self—giving the self identity—requires acknowledgment. I acknowledge my "self." I acknowledge that "the self" must be honored, must be nurtured, must be successful, and must come as a prerequisite before all else. All considerations of the self must be above all else. You have to actually claim the self, claim that *you* are important *you, yourself.* Ask yourself, "What am I doing for myself today, this moment, this very second?" Our selves get lost; our selves get betrayed; our selves get hurt every day. We must Band-Aid that self immediately or she'll go off somewhere else to repair. We will then become "soul-damaged." The term "soul damage" is the recognition of a lost part; we must retrieve it and bring it back home for our immortality.

Let yourself experience your unique relationship to yourself in your absolute surrender to you. You must always aspire to the sense or surrender, because without *surrender*, there can be no real identity at core soul level. You must try and merge with your soul identity through surrender. The aspect of surrender is a complete letting go of all there is; it is the knowledge that you can realize your freedom and identity through allowing yourself to believe in your *truth*. Allowing the truth to be really established as part of your core identity will bring much life force and real magic to you at your soul level. You are returning to yourself, and you need to feel that this returning of yourself to yourself brings to you all that you desire for your journey ahead. Just feel that your soul, or core self, is strongly connected to your truth and your ability to know exactly where it is in you that needs balancing and clearing. Surrender to this core soul part of yourself is the essence of embracing the concept of the Immortal Woman.

Merging and mending your damaged parts just creates in you your belief in your uniqueness and wonder of yourself. To be in wonder and adoration of the self requires much vigilance and trusting in the surrendering process. Just being in this space is a creative immortal experience.

The joy of surrendering is in knowing that, finally, the struggle to just keep understanding *the why* of living is over. The surrender takes you to another place within yourself where struggle ceases and joy and trust emerge. The joy of knowing you are a unique, love-filled human makes your struggle a joy in itself. This knowing to you is new; the real you is not somewhere else and brings to you your immortality.

Surrendering

The Immortals have the answers. We don't. We are humans. How can we have answers when we are experiencing a journey? We are looking, feeling, tasting, and being part of the experience of being human. When we find ourselves overwhelmed with all that is happening around us, we must stop—just stop. We must do nothing but go within and observe the self. The self must then make a decision to surrender the struggle of being human for just a moment. I don't understand this. It's too much for me, allows the surrendering to take place. And when it does, bingo! An alchemy—a knowing—takes place. This is the most powerful and loving thing that can happen to a human: the knowing, the love, the respect for the self. This self must trust itself to stop and really acknowledge that it doesn't have all the answers—no, no *answers* just the Goddess, just the Immortal Woman.

Emergence

The energy of emergence will come upon you once you recognize your truth. You need to finally allow yourself the freedom of just acknowledging that you are allowed to find within yourself a space for this light and energy to unfold into you and that your trust for yourself is above all considerations in making the journey. You need to feel the freedom of just allowing your soul self to connect with the concept of immortality; it knows it anyway. So you have to just convince the subconscious mind of the concept. The subconscious mind will pull out every excuse you and your culture, family, and past fears have fed it, but it can be trained. It has two sides, just like a coin; it can be flipped over, but the flipping takes effort, and this can be achieved.

Loving your total self enough to do this is what evolving is all about, so you must love yourself — your total self — enough to realize this truth and completeness. I am complete in myself enough to genuinely love myself enough to just grow and feel the pleasure of being in the presence of my immortal being. Being in the presence of my immortal being allows the truth to emerge, and my subconscious mind has no barrier to erect to create in me my resistance to my immortality. Just recognize that you can do this whenever the subconscious creates a barrier — now!

Being Present

The aliveness you feel at being immortal keeps you in the present, and every present moment has its own special magic. This is the joy of the status of immortality. It keeps you in the *present*. Being kept in the present is just a joy — the joy of being a human being who transcends her everyday reality by being immortal. To love your transcendent state is just so empowering because you know there is another part of you just waiting out there. The real you is also part of the out there. The beauty and joy of not only having to identify with your physical body and your physical world brings this amazing knowledge that death itself, in its ending, cannot really claim you. People are just so fearful of losing all that they have identified with throughout their lifetimes. Their death is the biggest everyday nightmare. Everything for nothing—just to leave *this*, all they have created, to someone else horrifies and appalls humans. *Yet it is for these very feelings you incarnate to become human.* It is the very thing you fear, that you transcend the moment you transcend the fear, you have taken yourself to another part of yourself. So, you ask yourself, "Am I ready for this? How can I do this?" This questioning is essential because part of the subconscious has allowed this thought to escape its critical nature in keeping you tied to reality.

To begin to look at yourself as immortal is to see yourself energetically as a woman who can tap into any reality at any given time. Your quest for immortality can have you viewing your past, present, and future realities simultaneously. The creative cycle of reviewing yourself and renewing yourself is a reinvention of yourself at every single moment, and you have to be fully open to this. Just allowing yourself to be fully open to the experience of renewing, reinventing, and reclaiming yourself is the joy and delight in experiencing the human journey.

Joy

The joy of the human journey is just that—the joy. So often humans forget that to be joyful can give them a mirror or insight into the experience of immortality. A truly joyful person is creating. She is creating life force, love, and truth for herself. Her life is journey and experience of creativity. Just allow yourself the joy of being human. There is simple joy in being human. Begin this journey into joyfulness now. The joyfulness of the journey is in acknowledging to yourself that you are allowed to surrender control over yourself.

I am allowed to surrender control over myself now. I am joyful in my humanness, and I create in my humanness as well. I surrender lovingly to the experience of being human by joyful living; this opens the door to immortality.

Peace

All we want to do as humans is to experience peace; *all* humans want is a peace-filled existence, free of the restrictions of time and place. Peace offers freedom from places and spaces that restrict flow. Peace gives space to grow the soul. The soul self is essentially one of gentle acceptance of the here and now. Your soul self knows that your world is safe; you are safe within it. Your world is your womb, and you need to feel safe in your womb, your world. Peace brings growth; when the soul is filled with peace, it grows and can expand. Look at a child. The child is filled with peace in its secure world, and it grows up loving itself. "There is nothing wrong with me," it says. The experience of peace must be sought if you are transcending everyday reality and developing immortal status. There is a huge call for the soul to develop this. There is a huge need to allow the soul to come to terms with itself. You need to feel free to be able to allow the experience of all of life to be part of your present reality. The quest for immortality brings to you now the need to find the essence of the peace-filled state. The essence of the peace-filled state is now the experience of knowing you can achieve many different dimensions simultaneously. You are safe in doing this when you allow yourself peace and love. Peace and love begin to get you there.

Divine Right

To allow your own divinity to experience immortality is indeed a gift to yourself. Your gift to yourself is to say to yourself that no experience of immortality is mine. It is my divine right. My consciousness just needs to expand to incorporate all states of existence. I am able to now allow myself the consciousness to expand and experience all states of experiencing reality. You just have to encode this reality with the consciousness you already have and just allow this consciousness to expand. Just imagine a balloon being blown up. Just feel your consciousness expand now. My consciousness expands to include multidimensional reality. This is the reality of being immortal. There is a new reality being presented to my existing view of life, and I am now going to explore the idea of just letting the flow of this new totality — this new experience — come to me now. I am not afraid to explore this totality now. When I do, I am going to be presented with a new, radical view of what it is like to be human. Being human for me implies trust in the experience of just surrendering to the multidimensional flow of just being.

All of life is a process now for me to experience. I just now connect to the divinity of myself. I am a divine human being allowing myself the feeling of space and love to be immortal.

Being in the presence of those who cherish you is an important admission of your need for self-worth. Your self-worth is an admission of your need to create in your world your worth. Your worth is the acknowledgment of your worthiness, and your acknowledgment that you need to be treated properly. Just knowing that you are cherished is a very powerful statement of your truth and worthiness. So ask yourself when you are not cherished why you allow this abuse, which causes restriction and no flow. To be unable to see why you are now allowing freedom and self-love to be missing in your life is your own self-abuse. No person can abuse you. It is yourself, which needs to be aligned with your truth to bring into your world your love of yourself.

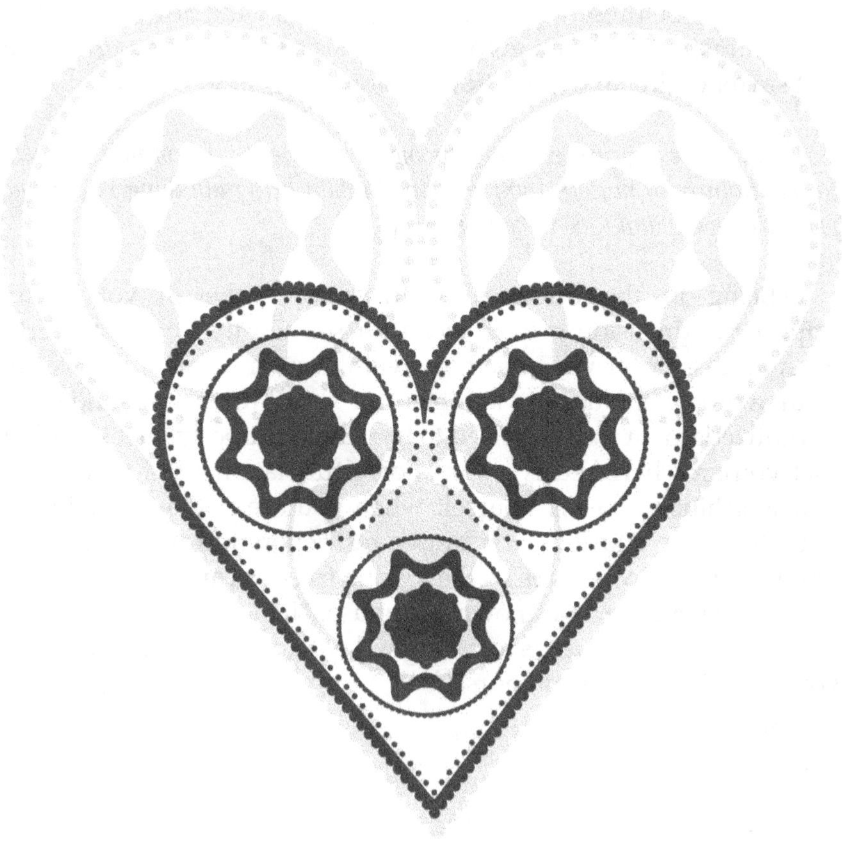

Chapter 8

THE SELF

True beauty and acknowledgment of the self now brings to you a knowledge that all of life is a shape-shifting experience. We shape-shift when we allow ourselves to experience life as a fluid moving thing. Shape-shifting is that; just infuse yourself with the identity of your total potential at every moment, especially in meditation. Just meditate on the new person you see yourself as being and feel that you and she are one. This is the essence of what being immortal is—just shaping your identity to create a new, more powerful one. I am shaping my new identity now, and I am superimposing my new belief about myself on to others. This is my truth, and my trust in this truth feeds me with essence and life force and radiance of that new identity. Just know that you, she, it—whatever the entity or identity is—is strongly connected to your truth and life force and radiance. So feel the essence, life force, and radiance of your own new reality. *You need to feel your own potential on yourself now, so you can feel the essence of the true you* emerge and take shape; you have now shape-shifted, and you have created your new identity, and you will find you are transforming into your potential: a shape-shifting, loving, cherished person.

Compassionate Love

The presence of love always creates in a person a sense of transcendence and immortality. The loving embrace of an energy of great love itself carries with it all that you desire for your journey of completeness. Being one in the presence of great, compassionate love brings to your world all that you desire for your journey of self-love. The feeling of just knowing you are loved by so many and that your journey is one of trusting in the love process brings to you all that you desire for your life and world. The sense of pleasure in sharing love is a very important reminder of your truth and completeness. I allow myself to experience the joy and intimacy of love to bring me closer to my truth. The sacred alchemy is the breath of life, and the love you share with others always has its rewards. The energy of love and respect for yourself is one of absolute truth and delight in the process of renewing yourself through the energy frequencies of *great love. Beings of great love* will always surround you when you allow yourself the space to really get to know you. You and the breath and the life force of your own unique essence bring the beings of light to you. This love from them allows you to really know that you will give and receive. They need your acknowledgment of their love for you so they can give you more. Both they and you will allow a fusion of great love, peace, and respect for all of life. Just being in the presence of great love ensures this.

I am in the presence of great love, respect, and peace. Love enfolds me now. I grow; therefore, I become immortal. Love feeds immortality. Love creates the alchemy for immortality.

Creating the process for alchemy is just that—a process. The process of creating an alchemy is one of acknowledgment that a certain reality is obtainable and possible. All of life is a process of alchemy; it is the trusting of special forces assisting you at a given time. The process of giving is just that—giving, surrendering to the process of allowing yourself the peace and love of completeness and love you need to feel. That sacred alchemy is an everyday occurrence *when the acknowledgment of sacred beings is present in the process. Never do anything or create anything unless the sacred ones are evoked for transmuting one reality into another.* Truly allowing the reality of just allowing sacred presences to infuse your totality brings the magic and alchemy that the light ones bring. Just allow yourself now to be infused with this sacred presence, this sacred alchemy of trust and love.

I trust and love the sacred presence of the light ones to infuse my everyday view of myself so I can know, in love, that I am a sacred being in the presence of sacred beings, to assist in the unique process of sacred alchemy. The energy and love of these beings for me creates in me a knowing and trust that my world is safe and allows me to live in a way that allows spirit and matter to merge into one.

This is a time of awareness and knowing that you are able to carry a *sacred truth* and responsibility for all in your life right now. Your life is now one of absolute love for this sacred alchemy to take place. There is a merging and oneness with all; there is an energy of great beauty and light. All of life creates immortality when the special magic of your truth merges in oneness with the spirit to create immortality.

Creativity

The reality of your life is the knowing — recognizing that you are able to feel in yourself your own unique sense of knowing that you are able to carry light and hope for yourself and your world. You need to feel a respect for yourself and all you have created. Your creations create all that will bring to your world completeness. To be creative in every endeavor is a very special gift you have as a human being. This creativity and essence will allow much magic to happen in your life right now. Just infuse every cell in your body right now with the essence of your creativity and life force. The divinity of who you are, the absolute divinity, will allow your world to expand to take in new aspects of yourself. The need to find within yourself brings to your world all that you need for your journey of completeness.

The human experience of immortality now creates much magic and excitement. You are now allowing yourself the "space," life force, and energy to grow your truth. This truth gives you energy and even more life force. Life force grows when you feed your truth, and you feed your truth with your creativity. This creativity and excitement brings just that—a knowing that, as a human being, you are above the mundane and sordid aspects of being human. This creativity is your passport to immortality; this passport is your truth. Not only must you recognize your own truth, but you must be able to *communicate* it to others, because it is a sacred gift of trusting in who you are. The sacredness of your truth now brings to your world all that you need to journey toward immortality. The divinity of who you are and what you can create is now bringing to you your belief in your ability to be an Immortal Woman.

Identity

A strong belief in the divine—a belief in the divine ones—allows you to really open up your imagination to encompass this reality. This reality now allows a sense of peace and love to be part of who you are. You are a creation of your divinity. Your own divinity is immortal. The soul is immortal, but when it becomes encased in a body, it loses its essential identity; its core identity cannot be recognized in a body and mind, which does not have a link with the unconscious, that which lies behind the veil.

The veil must be opened; you need to recognize that this veil must be parted to reveal the unconscious contents of human existence. The unconscious aspects of human existence are just that; they lie hidden behind the veil because the soul cannot reveal its immortality when it is unable to express itself; the human mind will not open the veil to allow the soul to speak to you—the human who has incarnated to give the soul growth. In realizing its immortal aspects, the soul is immortal. The body isn't, so you need to reconcile.

One third of your identity believes in immortality anyway. It's allowing the mind the freedom to really explore its own immortality, which is one of the basic challenges of being human. The human must challenge her humanness and look to dreams, symbols, and synchronicities that part the veil in order to uncover the hidden depths of immortality. Just acknowledging this is a powerful statement of your truth, in living as a human.

Soul Nourishment

To part the veil implies that your soul is ready to be revealed; it must reveal itself at some point in its journey in human form, or it becomes fragmented and just disappears or withdraws. The soul must be acknowledged every day and be allowed to be on stage. Visualize your story, your life, now and say to your unconscious, your soul, exactly what part it has had in your journey. In your story, just really acknowledge your soul's input here. This is a time of just acknowledging where it fits in. Give your soul food, just as you give your body food and your emotions nourishment. This is a time of acknowledgment and allowing the soul to *shine* — to give it its *time* on stage. The soul needs acknowledgment; it asks you what have you done for it? Ask yourself, "What have I done to nourish my soul?"

The time is now to recognize your worth at soul level and nourish that aspect of yourself. The soul is on a pilgrimage — a sacred pilgrimage too — and you need to recognize this. Retreat, alone time, meditative time, and sacred journeys nourish the soul and allow it to be given its power to fast track your evolution.

My evolution is now being fast tracked when I acknowledge that I have a responsibility to recognize all parts of my totality and feed and nourish that totality.

Just being in a space of allowing your soul to grow and recognize its true worth is now going to give you all that you need for your journey. You are allowing the soul to grow, to be fed, and to be made whole by paying attention to it and giving it what you both need now on your journey into completeness. Acknowledging the soul creates balance, peace, and contentment in your life.

Your trusting of growing your soul is always going to be a challenge because it's human nature to create a distraction whenever there is new growth or an unpleasant task. This is to be expected and acknowledged; just acknowledging this is extremely powerful.

I just know that there will be a resistance to anything that will make me develop. The struggle is to recognize this first and create a bargaining scenario. Okay, so I will give 10 minutes a day to developing a relationship with my soul and to developing an understanding of immortality.

You have the rest of the day; you can give me as many distractions as you can. The ability to create space from distractions brings a discipline and an acknowledgment that the soul's journey is essential for a happy, balanced life. Those who are happy and balanced have encoded in themselves that space to grow their soul self. It is very important for their journey into light and immortality.

Honor

The journey of the soul is an expression of honoring the total self; the total self needs to know that all of its selves are honored, so there can be peace and harmony in your life. This is an expression of your truth and your divinity, just express to yourself now your acknowledgment that it is important to give yourself "soul time" and that growth of your total selves cannot be in balance until soul time is acknowledged. Just acknowledge her soul/your soul: you and she are one.

I honor myself enough to develop a relationship with my soul and, thereby, ensure its immortality. I am an Immortal Woman who brings nourishment to all my selves through acknowledging soul time. All time is soul time when it is spent in loving surrender to the self — the core entity.

The energy of love cannot be overestimated in the process of self-discovery; through the process of self-love, the soul has a chance to be heard, to listen. The soul is a channel for divine love through the Goddess of All Light, the source of "All Love." The source of "All Love" is in the acknowledgment of the process of allowing the sense of immortality to take root. The soul needs a space to take root in the physical body and the mind, so it is very important to give healing to your soul today.

I am healing my soul today; I am healing my soul now by just allowing myself to experience this transmission from the source of all life, the consciousness of the Divine Feminine. I am in a space now to divinely love my soul.

Do this now. Take five (only five) minutes in private communion with your soul. Just five minutes a day acknowledges your soul, so you can really allow the soul to speak to you. Just nourish her in you now through self-love and self-nurturing.

Believe in your own ability to trust your soul enough to give her love, and you will find peace, harmony, and integration. You will feel as though you have done something for her (for yourself). This feels good, but it is a challenge to many of us. We exercise (hopefully) and nourish our bodies; we nourish our emotions by having relationships that sustain and nurture us. We allow ourselves the opportunity to now give the soul *five minutes* to really be part of our true self. Just light an incense stick, get a candle, place a crystal at your heart, and nourish your soul. Just nourish her/you now, you must allow this to happen; and, of course, you must remember your breathing during this process. The breath connects the soul to you. It is the link, the gateway; the link is always the breath.

Allow yourself the delight of knowing that immortality is just that—immortality and you are not living for nothing. Major religions believe the spirit lives on, and the essence of religious traditions has at their core this identity with the soul as being an autonomous, immortal entity. Your trust in knowing you are immortal gives meaning to every act. Every act of creation is an acknowledgment of your immortal spirit. The immortal soul just loves this acknowledgment: all of love is an acknowledgment of wholeness, of spiritual oneness: an emergence in truth and power.

Believing in the essence of oneness of your soul as an Immortal identity feeds your total relationship with yourself. This is just that—a total mergence into oneness and love. Just knowing that you are able to allow the spiritual essence of yourself to complete the cycle of being human is an acknowledgment of this power of spiritual transcendence.

I love and honor myself enough to just allow oneness to merge with me now for my truth. I am an Immortal being having an experience in a human body with the emotional equivalent; the essence of my living experience now allows for this mergence to take place now: I am alive to the possibilities of duality in all of life, and my immortality is an acknowledgment of transcendence of spiritual mergence. Mergence, oneness, and peace are now mine when I embrace the totality of being human in my immortal spirituality.

Trusting in the nature that life is a change process and that you are a "changeling" is essential for your journey as a human. The human experience is a trusting in the mergence with the source of light, and you need to understand and surrender to this at every opportunity. See this surrender as challenging and fun, and a feeling of freedom will come with it. Surrender. The struggle is aging: just stop struggling; it is aging and very destructive to the belief that you are an Immortal Woman. Believing in your immortality just allows you to really let go to the mergence and oneness of all. The Goddess of All Light is a light being of unimaginable strength and power and is a light source you can trust in the process of change. You are a changeling every day you surrender to source and trust the Source of All Light through the Goddess of All Light.

Changeling

Believe in your own ability to really challenge yourself here. Faith is an act of discovery of the self; just discovering the self is a challenge in your mergence into oneness, of peace where all desire goes, just allowing you, the *changeling* to recognize that every moment is a chance to surrender a previously held belief about yourself.

I challenge all previously held beliefs about myself, and I allow the source of all light to be my trusted guide. I trust the Source of All Light to be with me now, and I release previously held expectations about myself, allowing myself to be free. I am totally free now because I am a "changeling."

Why does a human want to be immortal anyway? Why do you as a human want to embrace the sacred philosophy of immortality? It is because you are listening to your soul; you are acknowledging that your soul has a space in your totality. To even acknowledge that you, a human, has a soul is a challenge for some humans. You can't see it or touch it; it doesn't have an identity; it doesn't have a personality. It doesn't have to be acknowledged at all really. Ask yourself, "Does the soul exist at all? And if it does, what does it want from me?"

First, acknowledging that your soul is a viable entity and part of you is a challenge for many people. Too often it has been confused with religion and tied up with religious practices and observance, which terrify many people. So you need to just acknowledge that there is a part of yourself that may have been "hidden in the cupboard" and kept out of sight for a long time. But you cannot separate from your "soul," for it is part of you, and that part needs acknowledgment.

Acknowledgment

By acknowledging the soul, you are acknowledging that you are whole and that being an Immortal Woman is just that — the belief and knowledge that, indeed, somewhere inside you, an aspect of yourself seeks acknowledgment. Acknowledging your soul gives you completeness, peace, and love. Living this way embraces the totality of yourself, and this is essential for a fully integrated life. Acknowledge your soul in love, and you allow yourself the freedom to be an Immortal.

Being an Immortal is acknowledging that you are part of the big picture of creation, not just a tiny speck of dust in the cosmos. The acknowledgment of your immortality leaves you with a sense of control over your life now. Your life leaves you with the opportunity to allow the beings of love to channel their divine love to you, and you can forward your love for yourself by acknowledging that you are in love with yourself enough to create in your life your divine plan; you are allowing such co-creation.

It is an exciting time to allow spirit to merge into matter here, and that you are able to be part of spirit. Immortality gives you power over everything in your life including adversity and tragedy; it allows you to connect with the overall plan of the cosmos. It is all right at that moment in time because the connection of the soul is through the process of immortality. Just feel and think of the divine plan in everything; everything has a divine order. The divinity and plan of your life is now allowing you to co-create with all of life and all there is: there is a true surrender and trust in just disconnecting from everyday reality to the divine reality. All realities have a master plan, and the master plan is one you can tap into now if you want to connect with all of life.

The essence and consciousness of the Immortal Woman is the acknowledgment that your life will unfold at exactly the right moment to give you *exactly* what you need at any given time. All time is present time, and you must allow yourself to see yourself unfolding like a lotus flower at exactly the right moment. Every experience is an unfolding of your true worth and respect for yourself. You need now to see that your life and all that you have created is a respect for the power and life force within you now, for true unfolding and emergence to take place. The need for acknowledgment of the trust process that everything will unfold and reveal itself at exactly the right moment will give you the opportunity to really be in the space of your divine present. The unfolding of the lotus is as precious as the full open flower. This is what trust is; it is the acknowledgment that you are able to emerge gradually, slowly, and perfectly. All of life is now emerging into oneness — oneness of peace, respect, and love.

I love, respect, and give myself peace at every opportunity to watch my emergence every day!

See a new growth; this is the magic of creation, of sacredness. Sacredness creates respect for every living thing in its moment of exquisite creation. The exquisiteness of all life is exactly that — exquisite in love, exquisite in its essence; all moments of creation are exquisite in themselves.

I am exquisite in my essence and creation to unfold now!

The essence and consciousness of yourself is the catalyst for your relationship with the quest for immortality. The immortal status is one you confer upon yourself every time you intone that the possibility of multidimensionality exists. The possibility of multidimensionality is just that — the possibility of an existence that has its links back into time.

Time Traveler

Immortality is allowing you to be your own time traveler. You can pick up snippets of your former aspects through rewinding your energy clock backward or forward in time. Try and not think about this in a linear, left-hemisphere way, but become a witness to yourself in this process. Try it with a person you have a strong like or dislike for, someone you have formed an energetic relationship with. You will find in your cellular memory traces of long-forgotten fragments of this early connection; just try and jog those cells in your memory banks. Just intone:

I ask for access to the source of life for my relationship with…

Getting to the source memory will track the energetic relationship back to its source; you will then have a background for the energy dynamics between you two. This will provide you with the information you will need for your journey together. This will prevent you from forming the same energetic patterns. The energetic pattern always needs examining when you are being presented with a new relationship; just allow the trust and surrender process to take place. The trust, the surrender, and the acknowledging are for your combined higher goods. Doing these things forwards the relationship and makes it an energetic one, where you can both move forward into *now* and the future time. Clear the energetic pattern from past relationships into the present and then create a new reality. You need to allow yourself the trust that multidimensionality will drive you on to uncharted parts of yourself that *have their roots in time long ago.* The source of all your personal power and all power is only that. Power over your own emotional reactions to life's events will come to you in abundance when you really allow yourself the "scope" multidimensionality brings. You need to feel the energy of love for yourself when you are exploring new, uncharted aspects of yourself, and you should feel that you are able to defer to yourself your wish for your happiness.

Complete Happiness

You are doing all these things to bring you *perfect and complete happiness,* so you need to express your love for yourself by offering yourself as a conduit for the energy frequencies that pour into you at this time. This is a time when you need to feel you can really take your own power into your own hands and allow yourself to release to your higher good; allowing life to be a complete experience is the essence of being human. You must trust in the experience of allowing yourself the trust to really enter into the experience of just allowing the Goddess of All Light to enter your energetic matrix and become a time traveler, allowing you to explore and experience multidimensionality forever. Trust in the process of forward time by bringing the possibility that it exists to yourself now!

Multidimensionality

Being in a state of oneness of reverence for yourself is allowing yourself the opportunity to be a multidimensional human. Multidimensionality is a process, a state of allowing yourself the unique opportunity to really be in the presence of the ascended ones, for you cannot be immortal without the knowledge that a state of multidimensionality exists. You are being true to the essence of yourself now. This is an important moment. You are defining yourself. I now co-create in trust and love my belief that I am multidimensional. To be immortal is to allow the process of multidimensionality to really be part of your energetic core identity. Your energetic core identity is a powerful statement of truth; you need to feel that you are able to offer yourself the love of yourself for this mergence and union to take place. The true mergence and identity creates a space of absolute love and devotion in you.

Core Identity

The Goddess of All Light merges in oneness with this core identity as she can connect with this love of self. She cannot connect with this love of self when the core identity is misshaped. She cannot look in her light, where there is no core identity, which has as its basis a belief that, as her "child," you will not accept immortality. She is in a space of loving acknowledgment, that "you" must allow the true trust process to take place. The true trust process is one of knowledge that you are in a space of oneness. Trust in yourself now to begin the journey. Just begin the journey of creation now.

Just allowing the true trust process to take place is an everyday opportunity to create the truth in yourself and your world. Your world will bring you much now when the essence of the truth and trust merge into oneness, just trusting your truth. Just acknowledge to yourself that the essence of yourself, the core essence, brings magic and power to you right now; there is a purity and absolute beauty in the true trust process. Your life will open up to an unbelievable degree when this sacred alchemy takes place. The space for this to take place is upon you now. You will find mergence, oneness, peace, and absolute beauty in everything that you do. The essence of the truth in the trust process acknowledges the special process, which is the essence of your unique oneness and absolute delight in the full process of life now. Just living life fully is the acknowledgment that the life force is merging with the full trust process. Just being in a state of oneness, grace, and beauty with all of creation is a truth that cannot be ignored. Just try to allow yourself the space and truth in the truth of the trust process. Be *in a space of surrender and knowledge that you need to find within yourself, at this time, truth.* There is an opportunity to really allow yourself the knowledge that life offers all that you wish in the truth of the trust process. Be in loving surrender to your immortality through the truth of the trust process now.

Communicate

Being able to communicate this trust process in your philosophy brings The Immortal Ones to you. The Immortal Ones are waiting for you all the time to you evoke their energy, life force, and presence to bring the multidimensional way of viewing life to you now. Just surrender to this love and trust that you are now in the presence of the sacred ones. The sacred ones now command that you be in their presence. The commanding is just that your wish for your understanding of the belief in immortality, actually commands them to you and this is how it should be, because they are willing to be at your command. You need to feel that the energetic presence of these energies is at your disposal if you command them in love for your journey. It is just that simple, but your energy frequency must be in alignment with your truth for this process to take place. There is a mergence and oneness. With this truth, you just have to allow this to take place now. You simply have to really trust the trust process because there is a divine law of trust, and this must be observed now.

Just trust the trust process, which is above everything else. Your life allows for this true trust process to take place now. You are in loving and divine surrender to this truth now. You are ready to really explore the truth through the trust process now. Your life opens you up now to absolute beauty and joy in the journey of trust. I allow this to be happening lovingly now. I command the trust process through the law of trust to be evoked for my life path right now to bring immortality to me.

Allowing the presence of the immortal ones brings to you the acknowledgment that you are able to be in the loving presence of all there is for your journey into the light. The light requires you to be releasing now all thoughts that are not compatible with this reality. The reality of your totality is just that—your reality. The reality of yourself creates in you the need to really embrace other realities. Just think of other dimensions of reality. Your "reality" is based on your past—your remembering of things past that your subconscious can tap into. You need to explore your concept of reality. What is real right now? What isn't real? You need to find what is real in your energetic makeup and just allow this light to keep coming down toward you. Just allow the light to keep spinning toward you so you can feel the powerful energetic link with your truth. The link with other realities imposes on you the need to really embrace the totality of reality. The true reality process, just surrendering to this concept, allows a new truth to emerge, and you will not be afraid to really allow the process of immortality to be one of your energetic makeup. You need to feel and understand this more: your truth is now upon you. Do you want to embark on reality that imposes a separate truth in the creation process? The creation process lets you really connect with all there is. The need to acknowledge this part of yourself creates a new dimension to your life. All of life now allows for the process of trust and oneness with all there is.

Absolute Surrender

Embracing yourself in your truth is now allowing you the pleasure surrender brings. There is a youth-filled energy in surrendering. *Just surrendering to your truth through the process of immortality allows a sense of absolute wonder in all of life.* Life is a wonder-filled process, one of absolute trust in the truth process, just allowing the wonderment of life to really give you all you need now allows you to partake in the mature reality at its most creative level. You are able to create and just allow this creation process to be really part of your energetic makeup when you embark on the journey of personal immortality. The journey of personal immortality imposes such a commitment to a shared sacred truth. Your truth becomes one with The Immortals, and The Immortal Ones co-join with you in understanding the nature of reality at a multidimensional level. You cannot begin to understand this just now because your view of life just doesn't allow for such shared truth to become your truth. Your truth is not your own necessarily; it is a shared truth. It is a sacred, shared truth, one of absolute surrender in the creation process. The creation process is just that: creation—the creation of oneness and a merging with all there is. Just allowing this sacred, shared truth to become your *one* truth will create in you the absolute truth; the divine ones are ready to show you.

Intent

The beginning of your journey begins with your intent; the intent is behind all outcomes so you must *be* clear in your intent. Your intent now is to begin to allow yourself space to create the intention: I now lovingly allow myself space to create the intention in love, life, and wholeness. The love of your intent must be felt and nurtured. Grow your intent now. My intent is to bring the belief that immortality exists and to give myself peace, love, and happiness in my life. I intend now to be in a space of surrender to the intent process. The intent process involves allowing the lesson of my divinity to be made manifest. This is a time to create, in this sacred intent, my truth. Both my truth and the energy and essence of the beings who are immortal will allow the intent process to be fed and nurtured every day. The intent process is the hardest thing for the mortal mind to come to terms with. Just allow yourself *the intent;* this creates a huge shift in the cellular makeup of the human brain. All consciousness is at the brain stem. Consciousness stems from this primordial patterning. All patterning must be recoded with new intentions to create a new reality. I am coding myself with a new reality through my intent. The essence of my intent must be now lovingly nurtured and supported. I am a product of my intent, and this intent will bring me all that the true process of immortality allows. Believe in your ability to create intent now for the immortality of your "self."

The cellular structure responds to this intent and is in turn feeding all other systems in your energetic makeup: your intent creates a new reality and a new belief in the process of creation; the biggest creation is yourself. You are the biggest challenge: no job; no relationship; no concepts, ideas, or beliefs hold the challenge that the self demands in order to grow. The self is like the most powerful light switch: why "not" turn it on to really allow the dynamics of your life to take place: why not challenge yourself to be an Immortal Woman now. Just say, "The biggest challenge on this planet is not it or what may happen in the future but what I do *now* to create in *me* the challenge to really surrender and submit to my truth. I have to trust this challenge now. It is the most important thing. Nothing on this Earth is more beautiful than the humans who create in love *the challenge of their being immortal.* Just being in this loving presence will bring to you all that you require for your journey of true trust and surrender.

Challenge yourself to be the very best human: challenge yourself every day to grow a new aspect of yourself. I challenge myself *now* to embrace the true trust process in all of life and light; I am whole in the true trust process, and I allow myself this challenge now. I am unafraid of the true trust process this challenge involves. I am loved, and I feel loved in turn. Be in your truth now. Bathe in the love of the Goddess to bring to you the beautiful, amazing energy frequency you are capable of. Do allow yourself the privilege of really getting to your truth now.

Support

While the challenge to be immortal is one of shifting focus from community and shared cultural beliefs about yourself, you will be supported in this endeavor: allowing yourself this support is very important in your life. I allow myself to be supported in making this shift, and I allow myself to be fully aware of how I must be in a space of truth for this to take place. I honor myself and my belief about my life to create this view of my truth now. It is important that you fully allow this mergence to take place slowly over time and you now allow yourself the ability to believe and fully integrate this aspect of yourself. You are afraid of no one, and you need to be answerable to no one: you are developing a sacred, divine philosophy, a philosophy of absolute love and honesty *for* the self *by* the self. You need to begin to slowly allow the mergence to take place and know that this mergence will allow spirit to be made manifest. Trust that spirit will allow your truth to be revealed to you to bring to you all you want for your life to unfold in order to embrace this concept of being. The new totality is one of love for the self to really feel that you are capable of allowing such a concept to be part of your reality. Why not lift the barrier? Why not just surrender all previously held expectations about your truth? Why not allow yourself this wisdom of knowing, just knowing that your needs have to be met? Know that you are embarking on a new journey of trust in your truth and that the journey will bring the joy of surrendering to the Immortal Woman to you now.

True Trust

Being in balance in everything that you do is the most important requirement in manifesting this new reality, and to manifest it, requires patience and loving trust in yourself in the true surrender process. The process of true trust and surrender implies that you are really ready to embrace the most amazing truth your reality can encompass: the true trust process to embrace a philosophy will challenge your previously held core beliefs about yourself and the nature of reality in general. The core beliefs now leave you free to embrace the philosophy of self-love and self-nurturing in every possible aspect. The reality of this core belief now challenges you at every level to really allow the true trust process to take place within you. You are now allowing this true trust process to create in you your belief that all in your world is really a force for light, strength, and power. The radiance in recognizing your immortality and truly believing in it are now your truth. Your ability to accept and embrace this amazing philosophy brings to you all that you require for the journey embracing immortality. Just allow the true beauty and trust process to take place now. Your life is one of absolute surrender and true trust to believe in the magic of the divine philosophy you are carrying. You are now radiating such love, joy, and power that all of life just teams with life force. The beauty of surrendering to the divinely feminine energy will give you all that you require. Immortality means trusting and surrendering to the belief that *you*, the *soul*, are/is immortal. "I am an Immortal Woman," is your soul's food. You feed your soul this way.

Beauty and Magic

The beauty and magic of now is just that: beautiful and magical. You are truly beautiful and magical when you embrace the true trust process to become immortal, and now you need to allow the beauty of all of life to *embrace you*; you are just too wonderful when you fully embrace the philosophy of immortality, and the truth of your life finds you truly open to the experience of oneness. The delight in the true trust process is one of total surrender to all there is. Your relationship with your truth is now one of the true trust process in all its aspects. I am alive to this true trust process embracing the philosophy of oneness the Immortal Woman embraces; just allowing the true trust process to be one with you allows the true you to emerge, and you will find a complete sense of peace and the knowledge that you are now embraced by the Immortals themselves. The immortals are waiting for you to really become one with them in the process of oneness. Allow the process of oneness to really make you alive to the power of now. The Immortal Woman is a belief in your authenticity — in your every moment. The moment of your truth is upon you now to fully support yourself to create the most outstanding relationship with yourself and all there is. This is a time of spiritual maturity and self-responsibility to trust. It takes spiritual maturity to really *trust* in all there is to bring you home to you.

Self Responsibility

If you are choosing the path of immortality as a personal philosophy, you will need to take absolute and complete responsibility for everything that happens in your life. This part is the toughest part of your human journey because, everywhere, you are being presented with options to lay the blame on someone else; it is just so easy to do this. My bad marriage, my shocking childhood, the circumstances of my birth, and so on and so forth—this must stop if you want to challenge yourself to a fully realized path. The energy and essence of a fully realized path creates in you devotion and loyalty to yourself. You are a living piece of artwork. Imagine the most stunning sculpture; really sculpt yourself on yourself now. You are a timeless treasure in human form. However, for this to happen, you must visualize yourself releasing all the negativity that makes up your life—all the stored angers, hurts, badness, and rage—and just sweep them all away. Your responsibility to yourself will grow, and every time you don't (or choose not to) hold on to a hurt, you are refining yourself and creating the perfect, sculpted shape: really create yourself now. Really allow yourself the picture of true beauty. The truly beautiful artwork is you in your true essence and love for yourself. This time is one of pure peace, love, and light for who you are. I just absolutely love who I am now, and I bring to my life all that this love creates. This love creates in me now a sense of wonderment in the true trust process. The true trust process is taking full responsibility for your creation: you. You have created you; now you must feel free to release that creation and be a witness to your wholeness: be a witness to your wholeness now!

To be embraced by the consciousness of immortality is to live a fully conscious life because you know there is no end to anything; everything is cyclical and has a pattern of loving trust. Just feeling the loving trust in everything makes all the effort worthwhile. You need to feel the total surrender to the struggle of anything that has to have a deadline. The structure of immortality cannot have a deadline; it is not linear; it is cyclical. The cyclical nature is a progression and evolution, and you need to feel the progression and evolution of your journey, without having to have a destination, like a bus stop.

There is no stop with evolution. The concept of immortality does not stop; you do not get off at the destination and go somewhere else. You are now ready to really embrace the cyclical nature of reality.

Evolving Spirit

There are only concentric circles; you just keep spiraling outward to ever more and more light. Your imagination just keeps getting more and more open; you are opening imaginatively, and you need to feel that the concept of immortality is just that—an imaginative journey outward. This journey outward is an evolution of your spirit, and you need to just feel that you are in the presence of the immortal ones, at every moment, as you know your life takes on new meaning and wholeness. To embrace this new way of being is to step outside the mainstream. To step outside the mainstream, brings to your life a richness and depth because life is not linear; it is cyclical. To now be alive—just to be alive—is to be immortal. Now to really be alive is to allow the spirit of the immortals to be with you at all times. The spirit of the immortals is comprised of all those beings who have lived consciously and in communion with the light of spirit in them since the beginning of time. To be part of the joy and magic of creation, to be in the company of the immortals, implies that your life is one of just accepting and realigning to this magical and awesome energy. The joy and expression of your truth is a testament to the spirit of oneness and creation for all there is.

Embrace

It is just so easy to find a space within yourself to be in the magical embrace of the immortals. Just imagine now all the immortals around you applauding you for trying to release yourself from your humanness with all its petty concerns and just embracing the totality and completeness an Immortal life gives. You are being birthed to this level of consciousness now, and you must feel free to really embrace the totality of your immortality in your truth now. The rewards are just so great! They include the ability to know that you are protected by the immortal ones, they who have transcended the limitations of being enclosed or enslaved by every emotion that tends to suck you back into the sludge and mud of humanity. Every emotion that is not surrounded by light and protected by the immortals has a way of dragging you downward, spiraling down and not up, so allow yourself to trust the light spirit of the immortals now!

Cellular Renewal

Infuse your cellular memory now with the concept. Believe in your immortality now by saying to yourself that you command your cellular memory to be infused with new frequencies. The new frequencies create a space for the cellular reconstruction to take place. Every cell then responds to the challenge. New energy is placed inside it. You need to find focus and direction, peace, space, and light for new frequencies to be encoded on the cellular makeup; the patterning allows for this to take place now.

I am in a space now where I am challenging myself to draw to me all the patterning, which creates aliveness in me. The aliveness this brings infuses my very being with the most incredible love-filled energy, one of great power and strength. Just try to imagine now your cellular memory being recoded to accept new frequencies—new information to allow the state of aging as we know aging to be gone; this is a time of allowing the peace of spiritual love and energy to really create all that you want for your life of love, peace, hope, and beauty. The philosophy of immortality gives hope and presence to everything.

There is a hope and presence and beauty to everything. Just bring that hope, beauty, and presence to everything you do right now. To infuse the cellular memory with the patterning you need to bring the essence and totality of all you have created in your life right now. There is a need to find within yourself a commitment to a way of being that encompasses immortality. The essence energy, life force, love, and truth of your truth makes your ability to really trust your cells to respond is creating all that you desire.

To bring your cellular memory into resonance with your truth requires courage and vigilance and, above all else, the knowledge that you are actually above all else—the knowledge that you are actually able to do this. It is important to allow yourself time to give yourself space to really feel this new direction and truth for all you are creating. Your life is now one of just allowing and merging with your truth at every turn. Life is a rich tapestry, and you only have to be aware of your truth in honesty and integrity for yourself. To rearrange your cellular memory requires that you restate your truth and make it manifest.

The Heart

Your truth comes from your heart, and your truth is always in the heart, so listen to your heart and feel your heart as a living, pulsating, loving friend. Your heart is your truth; your heart is a living, loving energy, and your truth is now going to allow you to really find that the tapestry is able to be woven with such beauty, intricateness, and complexity. Try to create all that *you* need for your journey on the road of life by staying close to your heart, so you can restructure your cellular memory to re-create new frequencies. This is a time of just allowing this emergence to take place. Life is a storehouse of unbelievable treasures and experiences. To be immortal is to really tap into this and know you are able to be free. To be free implies that you have choice, and it is very important that you honor your choices. This is a time of great expansion in your path because you are challenging a previously held belief system; you are challenging old pathways, worn-out roads, and creating a rich, new tapestry for your life as your cellular memory is recoded.

By staying in the present and close to your heart, you are able to find within yourself the true trust process to bring the vibration of trust to you. There are sacred laws, many of them in fact, and you can evoke the sacred law of true trust to help you redefine your identity for true trusting.

I now allow myself to experience the pleasure of true trusting in all there is to bring me peace, knowledge, and definition of spiritual laws.

Spiritual Laws

Spiritual laws are important to recognize and work with. In your emerging identity of wholeness with the status of immortality every day, you can evoke the practice of truly trusting the universal sacred law of true trust.

I truly trust the sacred law to bring to me the surrender required to truly trust in immortality.

My definition of myself does not include this; however, the true trust process is now being evoked for me to challenge my previously held definitions of self. This true trust process is one of just surrendering to the law of true trust.

I now breathe in true trust. I now swallow the belief that true trust even exists, and I allow myself this surrender. It is important that I allow myself the experience of true trust, and I have no fear that I will not be able to surrender completely to this process of true trust. It is allowable for my reality to include the true trust process to enable me to find the belief and nature of immortality.

Being alive every moment, every second, brings the immortal ones to you. You are embracing the immortal ones with your own divinity and power and love of yourself now. It is a most rewarding journey and one that will offer, in its own way, instant gratification because you are seeing goodness in every living thing. This, in itself, is a profound link with the divine ones and will create the sacred alchemy of trust in all of life. All of life is a sacred trust in the joy, belief, and completeness of everything. The joy, beauty, and completeness will bring spirit into matter, and spirit into matter creates sacred alchemy. The true trust process allows for such processing to take place. All is oneness; all is merging with the truth of the goodness of all light. There is a sacredness and belief in the oneness of all.

All and you are one; all and you are the true alchemy of divinity.

The true trust process is this alchemy of one. I and mine are one. I am truth made manifest. I am the Goddess of All Light of life. The Goddess of All Light is me. The Goddess of All Light to be discovered lies in me. I discover new things about myself every day, and I create in this oneness and *all of life out of me now. Life and I are one; one and life are one. All is one; all is truth.*

The tradition of truth must be observed; you are observing this sacred truth when you discover *you*; trust that it is *you* which is the choice. You are the choice; you are the one who creates all. You are the *one*.

Spiritual Light

Being in a space of complete peace within oneself is an initiation for the Immortal ones. When you really partake in your renewal and immerse yourself in the light of spirit and trust, this peace is further embraced, so you create a bigger circle of peace. The immortal ones know the struggle of being human and applaud the soul's challenge to constantly recode itself onto human experience. The human experience is one of learning the lesson of *trust* and *surrender* to the Source of All Light, and the need to follow this light must be infused in every possible way. Just surrendering to light and being in this sacredness every day creates the peace and space to journey. Allow your life now to be one of absolute joy and surrender in this truth process, and it will bring to you all that you require now to take on the challenge of being human. To be human is to find the light in everything and not fall into darkness and despair. The truth of your life is just allowing your spirituality to be part of everything. Everything is a spiritual opportunity to create just a little bit more light in your life and the lives of all; creating light and peace will bring to your world every imaginable joy. The joy of being human is having trust and enjoying this exploration. The joy of being human is simply being in this sacred, divine presence.

Soul Expansion

The immortal soul keeps growing, and you recognize that it is you creating the space of oneness when you spend time on yourself every day. Every time you do something for yourself in consciousness, you create in you your soul expansion, and an expanded soul is one of great and profound excitement and love for all there is. Feeling the freedom and love of an expanded soul is one of great trust that you and it are one. The space of love, trust, and peace infuses you at every opportunity to create oneness and mergence with all there is. Just feel the emergence of your soul and gently water it in love and trust for you/her; this will bring to you a feast of love at every turn. An Immortal Woman sees the soul as a cherished, loved friend, one who is to be fully trusted in love and truth. This is one time to bring to you all that you need for your journey together.

Just receive the soul's truth now, and know that you/she are one in spirit. The joy of mergence with your soul begins with just acknowledging that you/she need total fulfillment in every way imaginable. The time is now to trust you and her to be totally one.

The knowledge that a human must look after her soul now needs acknowledgment; the space for this mergence can take place in any area of your life. You need to trust this mergence now and *ask yourself if you are ready for this journey of trust together to become immortal.*

Your journey is one of trusting that your immortal spirit is being fed and nourished every time you create a sense of specialness around yourself and your growth and achievements.

You need to focus now on creating the specialness, the magic, the aura around you and feel that sense of exclusivity and specialness for yourself because immortality is the creation of this amazing jewel for all of creation — *you.* You need now to create in your life the amazing beauty of the creation: you are your creation and you are the most stunning jewel.

Magical Jewels

What jewel are you now? Think of a magical jewel and imagine its loving creation from its journey into the Earth itself. The journey of a jewel is the magic of all we started in the Earth: the matter. We were part of the elements first, then our parents' elements. We were lovingly nurtured to grow, and we now need to make our facets — our brilliance — shine. Keep shining and polishing your facets; make them shine; make them grow; make them all one in love.

Which part of me — in other words, my emotions or my feelings — need polishing now? Keep polishing your facets and always remember to admire yourself. When was the last time you really admired yourself? Admire yourself; see yourself as a jewel now, and all that you do for yourself creates a further refinement of your beauty. Your beauty is your light. *My immortal soul is one of truth, and I surrender to it now.*

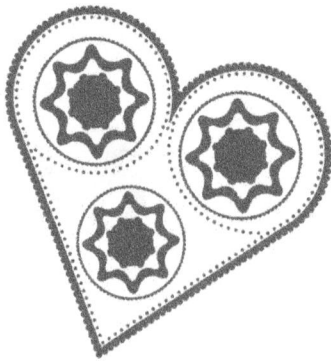

Chapter 9

SELF REALIZATION

To trust the process of simply allowing your life to unfold brings to you a unique opportunity to really connect with *all there is.* Just feeling the sense of true love and delight in your moments of self-nurturing will allow the guides to come. The guides are able to oversee the journey of your immortal soul and spirit. This is a time of just trusting in life and trusting in the process of self-love.

I love myself enough to bring to my life the sense of absolute delight in the process of renewal. This process of renewal will feed my immortal soul. I have an immortal soul, and it is fed by light. My immortal soul is one of absolute surrender to all there is, and I will experience a sense of deep peace and surrender. All I need to do now is trust the immortal soul to be of truth. The truth of my life is now one of sacred trust and surrender to all there is. This gift of knowing my truth brings all to me now. The energy of trust and truthfulness is for me to really enjoy. I am in a space of just bringing to my life all that is sacred, special, and wonderful: the true trust process is so magical that every moment in my life is one of special joy and magic. All is powerful and magical in the trust process. Being in this true trust process enables me now to fully partake in the joy of spirit and truth. All is love for me now in my quest for immortality.

I now accept the challenge of becoming an Immortal and ask the assistance of all immortals.

Ascension

The immortals just keep being drawn to a human who wants to ascend. The role of an Immortal is to teach the student how to create immortality in her consciousness. The consciousness to ascend is to grow your immortal soul to connect to the truth so you can be part of every living thing. An Immortal spirit is part of everything and is not separate. Separateness creates desire, greed, and hate, and an Immortal has purged these lower emotions from the psyche. It is important for the immortal soul to feel that every action while in a human body is one of completeness and trust; the trust and completeness rest with the soul in human form.

The soul in human form is one of allowing the knowledge and love from the immortals themselves to infuse every cell and every molecule of the psyche. It is so easy to become dross-filled in this negative, hate-filled environment. Being an Immortal is one of the biggest challenges in being human and one of the greatest challenges a human can set herself because of the need of the soul to be immortal in a human body. You can be an Immortal by just committing to the tasks outlined in this book and reading one page a day to infuse your cellular memory to accept new frequencies. All frequencies then are being reprogrammed, recoded, and realigned.

Just infuse your cellular memory with the new frequencies now by breathing the immortals to you now! The energy and life force received by the immortals infuses you with love, power, and respect for every living thing. This is a time of just allowing the trust between yourself and the immortals to develop. It is a time of acknowledging that you are choosing this particular way of relating to your world, and you must honor and respect yourself.

In this choosing, all is well for combined trust, love, and support. All is well for you to bring your truth to your life. The trials for this view of life bring to you many rewards, and you must feel free to develop your trust in yourself to master your new frequency. Just enjoy the spirit of this philosophy; it is challenging and demanding but most rewarding. Just feel the excitement and growth this new philosophy brings you. Challenge yourself in it and create the balance, peace, and trust in yourself now to be part of the most amazing light frequency imaginable. This is a time for honoring this truth, and you need to know now that you are able to be part of the process of life, which gives you answers to previously unanswered questions. Your life is one of hope, trust, and belief in your capacity to know you are fearless. In a fear-filled world, you are free when others are trapped, and you are awake when others are asleep; this is your immortal time, one of truth, love, and surrender to all.

Peace and Respect

Peace and respect for yourself is what the quest for immortality brings. Just create perfect peace by acknowledging the peace and respect for yourself. Release all previously held expectations of yourself and know that you are creating in your life all the desire for your life of trust and true truth. The divinity being conferred upon you will give you now the hope in life being one of absolute *embrace of all of life.* Trust truth — trust the truth of yourself — and because all of life is a trust/truth process, believe in the truth/ trust process now.

Allowing this trust/truth process to unfold reveals myriad mysteries. You may feel the need to create now the possibility of reaching out and connecting to the immortals intimately, exclusively, by intoning:

I am an Immortal Woman.

You are allowing your belief in mortality to be something that is not linear but multidimensional, and every cell is being reprogrammed too.

Believing and breathing light into your immortal spirit/soul brings to you the peace and space you need to process much emotion, which is being presented to your existence at a time of its evolution. The immortal spirit always lives and never dies; it is always ready to find renewal and replenishment in the true trust process. This is a time of just remembering that you are safe at this time and that all decisions will be ones of absolute truth in the true trust process. It is the time to *feel the energy that magical renewal brings to you.* Just intoning your belief in your immortality brings to you the connection with all that you desire for your creation into truth. A belief in the true trust process now brings to you a sense of magic, peace, and hope.

I am an Immortal soul having an experience in a human body. My immortal soul is my passport to my belief in my truth, and my immortal soul creates in me my truth. I am truth in my belief in immortality. I am peace in my immortality. I now find in my immortal soul my truth. I feel the freedom immortality brings. I feel the truth of immortality now!

The essence of an Immortal Woman is true trust that she is an Immortal soul having an experience in a human body, and the emotions, desires, fears, and beliefs are the challenge to embrace immortality. Every time you consciously embrace your immortal self you are embracing a belief in an answer to life's problems.

I trust that my immortal soul will be able to guide me in the pathway of being woman. I trust this truth now, and I am in a space of surrender to this truth.

I am an Immortal soul who has infinity as my home, is now stopping at one point in my existence and becoming human. I have decided to share in the human existence. I am aware of the challenges in being human, and I am aware of the pain, but I have made the choice to find an aspect of my total existence, which needs nourishment and replenishment in human form. There is an aspect of me that needs love, and in being human, I am challenging my immortal self to grow in this love. I am not afraid to experience being human, because the challenge in being human is to love. The single most important thing about the human experience is to be loving. To learn to be loving is the challenge and joy in being human. Just allowing this humanness to be embraced by my totality is my challenge as an Immortal Woman.

Self Love

The Immortal Woman now brings to her soul the experience of self-loving. The Immortal Woman is self-loving, so being human is a self-loving act. If every human could make self-loving a creative loving act, all would be well in the human's totality. Just allowing the creative act to be around you at this time is one of self-nourishment in the creative journey. The true creative journey is one of self/hood; the space of creativity is the challenge now to embrace all that "you" can in your life right now.

The Immortals are ascended beings who have no further need to be human, unless they challenge their new totality to assist humans at this time. They don't actually have to incarnate to do this, but they do need humans who are willing and trusting enough to surrender to their own truth to be a channel for immortality. Just allowing this to take place is happening now in human consciousness, and humans are responding as a group to this new view of their role in human form. It is important that humans have shared views on immortality now that they are not working alone or in a space of distrust. The world of the Immortals is a world of shared truth and trust, and just believing in immortality brings you very close to creating the spark that will bring the Immortals to you to work transcendence through you. You are able to allow this by a shared truth, a sacred truth, between yourself and others, and you can generate energy to bring this about. Generating energy to bring about your sacred truth is an act of love for yourself, and you need to respect the fact that you have decided to do this at this time of your soul's evolution. Just receiving the Immortals' energy is a very powerful and constructive thing to do for your life right now.

Surrender

Being in a state of surrender is a necessary requirement for the Immortals to connect with you. You can call upon the Immortals for everything that an ascending energy requires. Being in a space of great acceptance of yourself and living closely to nature and attending to the requirements of being human are very important. There is no point in wanting transcendence unless you are attending to the daily requirements of living a totally human life, living decently, and lovingly toward yourself. This is a time of just allowing all there is to surround you. This is a time of creating the peace and space to attend to you so you can be in a space of absolute trust in yourself to ascend lovingly. Ascending lovingly is a requirement of being human. Ascending lovingly brings you peace and love. Peace and love and surrender make a person whole. Just surrendering to the wholeness of yourself and knowing you are ready and capable spiritually to ascend will create the correct conditions for your truth to unfold. This is a time of shared love and truth and a time of knowing that you are capable of trusting in the light to bring you all you need. An immortal soul is a beautiful sight, and she radiates so much peace and light.

Immortal World

The immortal spirit of the soul is always trying to reach the subconscious in whatever way it can to allow the subconscious and ego space to find acceptance of this view of reality. Just try and remember why you are reading this and how the circumstances came about. The circumstances or coincidences are always the soul reminding you that you the Immortal Woman is being accessed through the spirit of those you meet. This is a time to accept this view of yourself and accept the reality of your immortal spirit—your immortal soul. The space for love, trust, and surrender brings to you all you need for your life to be one of peace, hope, and beauty: Just ask, "Where am I at in my immortal world?" Bring to yourself all you need for your life of truth and, above all, peace.

You will be shown a sense of absolute love and surrender to your immortal view of yourself. Believing in your capacity to transcend the limitations of your everyday reality shows you that you are able to create in your world all that you need to bring yourself a life of peace and beauty. Just show yourself that you are able to find within yourself all that you need for your journey of trust and truth now. Just find the sense of self through spirit.

The Immortal ones are alive, very alive, to the energy of Earth place, and seek to allow the immortal spirits of humans to connect with the immortals in power and love. This is a time in which the Immortals are alive to the life force of humanity's emerging consciousness. The emerging consciousness of humanity is attracting the Immortals in a way that is very exciting for your planet. Just allowing yourself to be receptive to this vibrant, new energy brings the Immortals closer to you. This is a time of emerging new beginnings for humanity, where the alignment of your truth and the truth of the Immortals is creating a new shift. When you are acknowledging the Immortals, you are connecting to a very important part of your cellular makeup, and you are able to ask for their assistance for everything that you need to evolve your soul's path: It's a time of joyful celebration of your immortality.

I now joyfully celebrate my immortality because I am free from the desire that my ego and subconscious conditioning brings. My immortal soul responds to the new frequency, and the soul is being balanced by the Immortals.

The Immortals speak of truth; the soul's truth and the soul must feel free to be really open to this love of itself now. The soul needs to feel freedom too, especially in a human body, as its nature is essentially free and unrestricted. The restrictions placed on the soul are ones of fear of change. All change is a response to the soul's call for its identity to be acknowledged in every way.

Your immortality is an acknowledgment of your truth and the belief in your ability to really transcend the limitations placed on you by being human. To be human and divine all in one is the quest of the immortal soul/spirit/body/emotional being. Just feeling your space and self-respect now is the true truth process. The space and respect of the true truth process acknowledges that the soul/spirit can be immortal. The immortality of yourself is the immortality of all those spiritually evolved beings who also want to transcend the limitations the human condition places on you. This is a time to acknowledge that the truth process is being set in place in your cellular structure right now. Defy the process of aging, rejuvenate yourself with the frequency of the Goddess, and bring to you all that your life requires because the process of cellular rejuvenation is about such challenges.

To challenge ourselves to be immortal is to feel the response of our true selves; a calling and inner need forward our true path. The truth of our path is the inner calling of spirit/god/goddess within. Don't be afraid of allowing yourself this inner calling. This is a time for allowing yourself to really find within yourself your belief and ability to bring to your life all that you need to celebrate the journey of being human. It is fun and an adventure to trust in the adventure because life is an adventure of trust in spirit. Trust in your spiritual growth for your journey of renewal now.

Awareness

This is a time of expansive awareness. Expanded awareness means that your spirit/soul self grows in the true nature of co-creation with the universal laws. The universal laws are simple, very simple. You only live to trust and surrender to the joy of your life. These are simple, joyful acts. Do them; trust them; then you are trusting us, your immortality co-creators. Just bring to your life this belief in your totality now. I am fully alive in the true trust process for all that happens in my life right now. This spirit of adventure is the source of true power over your totality. Just find the focus for this experience to be part of your life now: I am alive to the true trust of adventure for my journey now. Celebrate all you are and all you choose to create now.

Creating choice is what being an Immortal is: you have a choice. Practice making choices now. Say to yourself, *"I choose to create in my world my belief in my immortality."*

Slowly there is a gentle shift in awareness, and you now find yourself creating more peace and space in your life. The joy of an Immortal existence is that. Life never ends; it's never over. It allows the energy of the soul to fully partake in the human creation: the human creation is one of absolute bliss to all there is, and your creative self can fully partake in the journey of being human. Life — the life your soul created for you by acknowledging the source of all life, your life — allows you to really partake in the journey of being human. Just forgetting the recipe doesn't mean it is lost forever, and you need to feel confident to really allow your true soul/spirit self the chance to allow the connection to take place and occur.

Believing in your ability to trust the forces that shape your destiny brings to you now your immortality. Share your new belief about yourself only with trusted people, those who can understand your vision for your life: you need to explore your new philosophy with those who will support your identity and your truth. Sharing and believing in the true trust process will bring peace, space, and a desire for the true light to emerge in your world. It is time to find that you are a true creator here. Others will be able to share their philosophy about their beliefs too, and you will feel more confident in seeing yourself as multidimensional.

New Identity

Allowing yourself space for your new identity to grow implies trust in yourself in delivering a new life for yourself and being in a position to share this trust. First, trusting yourself is the beginning of embracing an Immortal life. Follow your dream for the true trust process to take place: Love yourself enough to give yourself this space and, you will allow your truth to emerge. This is a time of allowing the truth to emerge so you can not only evolve yourself but all those you love. This loving act is one of courage. We Immortals

CARMEL GLENANE

know that. We know what the soul must endure in human form, yet we do not allow your spirit to be dormant if you surrender to your truth. Just being in a space for your own love and emergence in the true trust process for yourself will bring the immortals even closer. Ask yourself daily, "What do I want for my life right now?" Then you will have the freedom and love the true trust process implies: just see yourself as developing such love and hope for yourself; this is the knowledge that you are loved and cherished by yourself. The knowledge that you are loved and cherished by yourself is one of great and profound truth. The truth of your life now is to embrace your desire for the true trust process to bring to you all that you need for your life of love and peace.

Try and observe exactly where you are when confronted with the challenge of self-change. Which aspects of yourself now are being challenged? Many previously held beliefs about yourself will be challenged in this journey of change. To challenge yourself now is to submit to the belief in yourself as a multidimensional being truly able to link with all there is. This is a time of allowing the spirit of your truth to emerge and take flight. Just allowing this is the new challenge—embrace it and your truth now. This is a time to really understand and respect yourself in your *divine* essence, your divine truth.

Meditation

Being in a state of simple surrender to meditation brings the Immortals to you now. Devote some time to your meditation on the Immortals, and just allow the spirit of the Immortal ones to be with you now. This is a time of just being in the true trust process for all of life. Try and surrender to this true trust process now and allow yourself peace, space, and light to connect to your divine, immortal truth.

I am allowing the Divine Immortal ones to come closer to me now, and I am not afraid to feel the Divine ones near me. I am forwarding my truth by allowing the Immortals to be with me now. This is a time of experiencing and expressing myself in my true immortal status.

Being an Immortal Woman is releasing and powerful as it allows the true trust process to merge with you and the Immortal one. I allow the Immortal one to be part of my reality, part of my consciousness now. This is a time of remembrance. Try and remember just who you are and what your belief system is. Really hold to this belief about yourself and bring to your life all that you need for your Immortal self. The Immortal self is now being activated at a cellular level, and you can really begin to respond to your new view of reality.

To resolve your inner fear and your inner guilt about your new reality implies that you are now making a shift to encompass a new reality and spirit. A new belief system brings this true trust to you now. I am not afraid to trust the true trust process *now* to bring the Immortals to me now.

The Immortals give me belief that my life is one of absolute trust and surrender to all there is and all there will be.

I now affirm my truth in all there is and all there will be. You are an Immortal Woman when you live by a philosophy that every living thing must be part of you and you are a part of every living thing.

We are all part of the one, and we cannot be parted by the one when we believe in the one. We are the *One*.

Allowing the spirit of yourself to truly embrace your immortality brings to your life your oneness with all that is sacred and life-affirming. To embrace this part of yourself is to really bring the joy of spirit and oneness to you now. You are an Immortal when you believe and trust in yourself to experience your uniqueness and wonder at all there is. This is a time for allowing this uniqueness and wonder to really become the one.

The One

The one is just that. You are part of the one of everything. To be one with everything is just so life-affirming, so good, and so whole. This is a time for creating the one in you now. This one is the belief in your total ability to really trust that the one and you are just that: one. This is a time of merging into oneness; to believe in your own ability to transmute the density and dross of your humanness is just so unbelievable to you that you cannot even go there with this energy. Just to relive the dross, misery, and fear of being human is the ability to really feel that you will bring to your life the absolute belief in your uniqueness and that the one — the Immortals — are listening to you. The challenge is to do this now and to create in yourself this energetic shift.

When you embrace your immortality, you are embracing the one of every living thing, and you are one with every living thing. This is such a massive change of consciousness that you cannot even believe that you, a human, could ever be so responsible for such a shift in consciousness. This shift in consciousness will create many exciting openings in your worldview right now — you'll be able to embrace the absolute one of all of life in all its immortal aspects.

To believe in the oneness of every living thing implies just that; the belief in the oneness of every, living thing. To really allow yourself this belief in the oneness allows you to connect intimately at every level. This level implies that your trust in yourself and others is really interconnected at every level. You are now bringing to your life this interconnectedness into everything where the true immortality of yourself can be realized. The Immortal Woman is you, and you are the Immortal Woman when you really trust in the true trust process in all of life. I trust in the true trust process in all of life now, and I believe the true trust process will give me all I need to grow and develop my immortal status. This is a time of true trust in all of life. All of life is mine. I am all of life. Just believing in the true trust process gives you all that you need for your life of true power and belief in yourself now. Just allow the true trust process to really bring to you all that you need for your life right now. Just keep intoning:

I believe in the true trust process in all of life.
All of life and I are one in this true trust process,
and I allow this to be taking place right now.

The immortals are the beings who offer more to humanity at this time. The Immortal ones bring hope and also challenge to humanity as well. This is important for humanity to nurse its hope in immortality because it not only elevates the human spirit but all of humanity as well. This time is now allowing more spiritual energies to be poured onto Earth. This is good for humanity at this time. Humanity must feel that this outpouring of love for your planet creates changes in the cellular memory frequency as well. This is the wellspring that feeds change, and this is the time of great change for you and all of humanity.

This change allows the beings of light to come closer to you. It is now a time for forgiveness of yourself and all there is to forgive, for your true mastery of your physical self to bring to your world not only hope for yourself but all those on your planet at this time. Just continue to support yourself in your daily endeavor to support your truth, and you will find that there is a true trust developing between yourself and the Immortals. You must trust the Immortals as well as trusting yourself. You must feel that you can trust every living thing, that you can surrender your fear of trusting the forgiveness of yourself and every living thing. Just remember you are loved and supported here.

True Trust

The true trust process has many aspects at many levels. The purpose of these levels is to really nourish yourself in your humanness and bring to your life the belief in all of life to you now. The belief in all of life creates conditions that give you the chance to really feed the life force energy and true trust that a consciously realized life brings. To create these conditions will now allow you to really embark on the creative cycle the immortal spirit brings. The immortal soul is a vehicle, a means of transport to have experience and ascend through the universe. Just allowing the expression of the soul to really experience this transcendence, this merging, creates uniqueness in the experience of being human. The conditions here create the sense of beauty and life force that an Immortal soul brings to all of life. This is a time of just allowing the true trust process to merge into oneness with all of life in the true creative aspect of being human. To bridge the gap, one needs to be in a receptive state and to find within him- or herself this merging and oneness with all of life. Allowing this aspect to merge and filter into your life now creates a sense of powerful challenge and opportunity to change your life.

All aspects of the self must be shared with the self. This is the essence of a happy life. The sharing of the self with the self, first creates the process of getting to know you. The getting to know you is an important act of being human. You cannot get to know you if you constantly share yourself with others. You are only revealing to them their/your unmet needs. Be very discriminating with others; it is as important as a soul getting to know itself through being in a human body and needing this space and time.

You should feel very safe getting to know yourself, as this is where immortality begins. The soul self now needs the challenge to meet all its needs through others. Just continue to nurture the soul self at every opportunity. When you are fearful of being alone or don't want to be alone, you need to question *why* you are so afraid of yourself. Why am I so fearful of me? Why am I frightened of me? Just recognizing your fear of yourself is very liberating. Why am I so afraid of me?

We are afraid of ourselves when we are not in our truth. We must not be afraid of ourselves because we cannot attain immortality unless the self is not afraid of itself. Create boundaries when you find yourself in fearful spaces within yourself and when you are being surrounded by fear in others. Yes, you need to recognize that the road to immortality, to be an Immortal, is to have no fear, no entrapments through affection, false loyalties, and attachments. Attachment creates the breeding ground for fear to settle. Fear settles where there is attachment, and fear is always around when the soul is enslaved by fear.

Fear and Evil

Your initiation is to really allow the fear that is around you to dissolve away. Feel it dissolve. Feel it just evaporate; feel it just be swallowed up by the Mother Earth herself. Feel it going down the plughole, just being eaten away by the Earth herself. In fact, it is good to embrace fear. It is good. It is pleasing to see the fear and know that *evil*, yes evil—the forces of darkness—have attached to that fear that attachment, that desire, that expectation. No expectations equals no fear, which equals no evil. Evil loves fear; evil feeds off fear; evil creates such a rampant, distrustful energy that it just loves fear. Give it fear, and you give it a breeding ground of absolute, magnificent, wonderful hate. The hate is the hatred of the self; it pits the self against the self, creating self-loathing: have none of it, and fight it with your immortality and the immortals now.

Truth

The immortal truth creates in you now the desire to really follow that truth and live it. Essentially, a life of truth is one where you are being bathed and infused by the light of the immortal ones. There is no greater truth than to live your immortal truth. There is no greater quest than to find safety in your immortal belief in yourself. You need to recognize that this truth is a powerful statement of your immortality. Immortality is a connection with your essential truth, and immortality is a belief in that truth. You are an Immortal when you trust and believe in your essential truth and live it wholly. The belief in your immortality is a testament to your truth and to your status as a truth bearer. This light of spirit infuses everything and everyone you meet. See yourself as being connected to this immortal spirit—this immortal truth—now by living a life of absolute truth in everything that you do, say, and feel. Our immortality places upon you the need to really infuse your spirit with your truth. Intone every day: *"I am an Immortal Woman. I am infusing my immortality on myself. I am Immortal when I believe in my true truth"*.

My true truth is for me now to really connect with, so I feed my immortality every day with my truth. Every thought, every act, every intention must be a statement of your truth.

Sacred Space

Creating a sacred space for your immortal self implies acknowledgment of that self. The self speaks when the Immortal Woman's space is honored. Your sacred, personal, and Immortal space grows from one tiny corner in your bedroom, living room, patio, or garden to your whole room, your entire home, or your whole garden. Intone: "This is my Immortal space, and it is where The Immortals are being nourished by me." Give a gift or offering to the Immortals so you can evoke their assistance for all you do.

Say and bring into your life; The Immortals create a space of great peace and reverence in your space and speak of you in a whole way. *My immortal self is my whole self. My whole immortal self is with me now in everything I do, say, and feel.*

When you create a "sacred" space, you honor your immortal self. Your Immortal self honors this immortal space, so the vibration lifts, and the Immortals can come in for personal rejuvenation, cellular healing, and transcendence.

You must remember that you are an Immortal having an experience in a human existence. It is not part of your totality. It is merely a space to experience the emotion of ... and to process, release, and, finally, to transmute karma. This is all that the experience of being human is.

The belief in the existence of yourself as a totality is a chance to really release those fears of abandonment and separation. How can you be abandoned when you are part of the totality? The part of the totality is being part of the human and nonhuman reality. Just feel and allow your heart to open to the gift you will receive as an Immortal. This is the gift of oneness, of sharing, and of being part of a worldview in which you see yourself as being totally and intimately connected to the source of love and power. This is a time for just being in the flow here and allowing this flow to merge with you and wash over you. Just be bathed in this love of great light and peace and create in your life all that you need for your life of forgiveness of self so that you can bring these unique circumstances about in your reality right now.

You are now going to bring to your world this magical and unforgettable experience of light infusion with the Immortals, who are the totality of who you are. The Immortals are the totality of who you are in every way imaginable. Begin to believe in this totality now, and you will have tapped into a part of yourself that will find peace, hope, and renewal every moment.

Intone, *"The Immortals are me now."*

Responsibilities

Creating in yourself your belief in the essence of your totality brings with it responsibilities. There is a responsibility to the self when one embraces this view of reality. There is responsibility to carry this truth for your life of trust in the truth process this philosophy brings. The Immortals create a great sense of personal responsibility for all that happens to you. You must check, observe, and note every emotion. This especially applies to *negative* emotions, which must be tracked to their source. All emotional reactions create barriers to your personal truth and, in turn, attract more problems for your soul's path.

Grief

The Immortals are always evoked when a human wants to create a space to release emotional reactions to life's challenges, which, of course, have a karmic pattern. The Immortals create a space

CARMEL GLENANE

within the psyche to observe and bring into focus all that you need for your expanded view of your truth. Embracing the philosophy of the Immortals brings to your life this expanded view of your life taking an overview of everything. Grief, for example, is a reaction to an emotional loss, whether it is a physical death or the release of a relationship or something you have had a strong emotional attachment to such as a pet, a piece of property like a car, or anything else you care about. Really examining grief and why you are grieving creates in you a chance to really embrace your immortality. Grief is a great opportunity to grow and experience life beyond emotional attachments.

In fact, grief is the most challenging aspect of embracing humanness. To be able to honor yourself in the grieving process and allow yourself to witness yourself in it creates the most powerful opportunity to give yourself your immortality. Your Immortal self grows through the grief process, as you are facing what you had and now don't have. Just observing the process and flowing with the pain allows your spirit to connect with you and assist you. All the Immortals are at the ready when grieving humans expose themselves to their fear and transmute it with love. The fear can be transmuted through the act of embracing the self and embracing the hurt.

Embracing the hurt and bringing it into yourself allows your immortal self to grow. Don't shut the pain away; just go into it; fall into it. Just give in to it and then explore the hurt of how you are feeling. People are afraid to really feel the pain. Just really embrace the pain: it cannot hurt you. Hurting yourself is in not embracing the pain. Feel it, really feel it in your heart, and then fall into it. Just fall into it, and just flow with it. Stay with it: don't be afraid to stay with it and then thank it. Thank you for what you are teaching me right now.

Being afraid of grief is a denial of your immortal self. Grief is merely attachment to someone or something. The attachment is part of being human. Accept that attachment is always part of the human condition, but attachment eventually creates grief because everything comes to an end sometime. Everything has a cycle, and grief is the

byproduct of attachment. I am attached to you; you leave me for someone else; you die, or I leave you. All of life is attachment. All relationships come from attachment when there is jealously, greed, or another emotion involved. This needs to be exposed, explored, and released. Grief is a byproduct of all attachments to outcomes in relationships.

Grieving is part of being human. A human knows that, but there is no way to show humans how to really prepare and come to terms with the end product of a spent relationship. When a relationship begins, it is inconceivable to the couple that they prepare for its eventual end. Yet, in preparing for its end and knowing nothing lasts forever, you are ensuring immortality for yourself and the relationship. Acknowledging the truth—I love you. You love me. We are together. It will end. You will leave. I will leave. One of us will die—creates the knowledge that every moment is precious and shared. Every loving act between the couple must be preserved and made special. This ensures Immortality. Your love then is an Immortal one because you haven't had attachment at all to yourself or the other person.

The Immortals are always creating in the human consciousness the desire to attain immortality because being human is such a transient experience. The Immortals are allowing us to come to terms with this by allowing us the freedom and space to get to know them in their true form. We can be an Immortal in human consciousness when we begin to allow ourselves the ability to truly surrender to ourselves. Our true selves are always at the ready to create in ourselves the true trust spirit of abandonment to our true relationship with ourselves. This is a time to really embrace our uniqueness. Our true riches of spirit are here for us. Just ask the Immortals to show you a glimpse of your truth and creativity. Guide yourself slowly on the path of immortality and don't be afraid to see death as the end of anything. It is merely the beginning of something new. This must be embraced and feared. This must be challenged. Always challenge loss; always challenge change. Embrace every day.

I embrace myself in my immortality now by challenging my truth and creating the spaciousness and expansiveness the truth gives me. I am alive to the beauty of my immortal life now; I am alive to the true trust process, which comes with immortality.

The true trust process, which must precede the quest for immortality, begins when you acknowledge to yourself that you must, in fact, truly trust. To truly trust is to surrender absolutely and totally to your truth. You will begin to find your life changing, and new people will begin to gravitate toward you. There will be a growing respect for who you are and a sense of allowing your immortal self to really emerge.

Ownership

The emergence of your immortal self gives you a sense of true worth and ownership of yourself as an individual. Take complete responsibility for all that happens to you. This ownership of our truth brings with it a sense of peace and delight in the true trust process. I delight in the true trust process, and I love the vibration of my immortal *self*. Waking up this true trust process is my gift to myself for it will give me the ability to see my true immortal self. My true immortal self creates the vibration of trust, and my true immortal self wakens to even greater truth. This truth is now going to create in me the absolute delight in the merging and creating of my "self," which is one of absolute abandonment to my self-love. I am now allowing my immortal self to create in me all that I need for my journey of truth. The essence of my essential self is in the transfer of energy through the process of immortality. To be truly immortal is to be alive to my true potential, as I can create all that I need for my journey of truth and love. Trust in the process of love now.

Spontaneity

The expression of your immortality can be in every action that finds you responding spontaneously; the spontaneous response is an Immortal one, one of surrender. Just surrendering creates immortality. The surrendering allows you to feel you can drop into something. Just fall into your belief that you are immortal. Just try it. Just fall into it. I now surrender to my immortality; I just surrender or fall into my immortal self. This surrendering creates in you a need to create a space for love and acceptance of your true, immortal self.

Surrender

You need to feel and find the freedom to surrender to your immortality now. Just surrender; just create a space of abandonment to your true immortal self. I am now creating my immortality by simply surrendering. Teach me to surrender *now* so I learn to surrender. I learn to surrender by just trusting that *I* can create the conditions to create surrender. I learn to observe myself as I surrender. I learn to create a space in my life for surrender. Draw a circle; now place yourself in the circle. This is my surrender space; I will practice creating my surrender space and pretending to fall into this space. Know your surrender space is a place you can go to be immortal. This is my surrender space; my immortal space is created through surrender. I am surrendering to my space of truth to create my immortality. I am creating immortality by surrendering.

Courage

Creating the surrendering space acknowledges that you are able to be part of the process of self trust. You must trust the space you are going to walk into. I trust the space I am walking into to surrender. To walk into a space to surrender creates the need for courage and again the word trust. You need the courage to trust in order to even walk into the space for surrender. So just do some deep breathing as you align yourself to the forces assisting you at this time; the forces assisting you will be at your side now as you trust yourself to be courageous. Trust yourself to be courageous now, and trust that the beings assisting you are really there; there will be a moment of fear, and this fear must be challenged. The fear has karma attached to it. The karma attached to the fear creates the barrier. Now, just feel where the karma is in your body right now; is it in the chest, the pelvis, the belly? Find the pain and send love to the pain.

Healing the Self

Try sending love to the pain now. Just try and release the karma attached to the pain by intoning to your light beings what you need for your life right now. Just create in you a sacred love for yourself as worthy. You are worthy as a person to go into your sacred space and create the trust in yourself that you need for your journey of truth right now, because your truth depends on this, my truth depends on this. You cannot evolve as a soul if you don't send love to that part of yourself that needs healing, because it is karma that is blocking the journey.

Begin to locate the pain and trust.

Beginning to locate the pain and trusting is the message now. Begin to feel the freedom of trusting and begin the journey of allowing the trust process to begin to unfold for your karma. Trust that the karma can be sourced and cleared. See yourself immersing yourself in a bath of pure light. You are now being immersed in that light, and you are slowly slipping underneath the silky, luminous "water"; its sensuousness and texture remind you of pure silk with the heaviness and sensuousness of cream. Now imagine that you have attendants around you gently bathing you like a mother would her infant.

Just experience this and bathe in its essence and purity right now. Your life, your immortal life, depends on this, and you are now really able to bring to your very being all that you want for your journey of truth. Trust your truth and never fear surrender because it is where true truth lies. True truth lies in your ability to really embrace your totality now and always. You now need to really feel that your life is one of true trust and truth to The Goddess of All Light. To allow yourself to be bathed in this gentle energy of the mother brings you back to the comfort and protection you enjoyed as a baby and child, whatever your upbringing may have been. All mothers' love will have fears attached to it because all mothers are human, and your journey must be to acknowledge that fear of your mother and go beyond it so you can transcend everyday mortality and become immortal.

Truth

When you allow the sacredness of your life to be your guide, your life is yours. It is a sacred experience and must be respected and nurtured. Your life is one of the truth. You must acknowledge the truth of your existence now, and you need to feel that you are able to express your truth at every opportunity. This opportunity is always there. Always be able to express your truth and always know you are free to be part of the mystery of creation, your creation. Your creation is your belief that all that happens to you is for a special purpose, which is your soul's evolution, of course. Just trust and believe in this now.

Honor exactly what you are doing in being human, *even the struggle*. Acknowledging the struggle puts you in a place of absolute power because you are admitting that you need help. Admitting that you need help is a most empowering thing. I am empowered in admitting that I am struggling at this moment and that I am in a position to trust that I can create all that I need for my life right now. Trusting in yourself to really embrace the challenge only reinforces the challenge to go all the way with yourself. You have to acknowledge at times, sometimes every day, that life is a challenge and that certain events are going to put you into this fear space. This is important to recognize and to go beyond. So be good to yourself and reward yourself for all you are creating now. An important time of growth is upon you right now as you acknowledge the challenge of being human.

The challenge of being human embraces all that can be reasonably achieved in one given area of life. For example, just acknowledging to yourself that you fully embrace the challenge of being human now, in the area of personal intimacy, is an acknowledgment of your willingness to surrender that part of yourself that is locked away secret and cannot be reached. You say to yourself, "There is a secret chamber here that I cannot reach that not even I allow myself to go into. I don't want to go there; it's too secret even for me to touch. It's too raw and vulnerable. Just let's lock it away and throw away the key." You must feel free to really trust that part of yourself and just unlock the cabinet. Today I unlock that secret part of myself that won't let me enter because it hasn't been healed. This is a very important aspect of transcendence. The very important part of transcendence is in going to that secret part of yourself and really experiencing you. You must feel free to walk into the cabinet or chamber and let yourself achieve intimacy with that secret part of yourself. You need to do this now and allow that secret part of yourself to heal. Open up the cabinet here and let yourself really be part of yourself; just open up yourself to yourself first and achieve true intimacy with the self: you. This is immortality because you can't evolve totally without being able to do this.

Commitment

Creating time in your life to be immortal takes discipline and commitment. You might like to challenge yourself now to commit to a period during your day to stay entirely "present" and be aligned to your true trust. To be aligned to your true trust requires the discipline to believe that you are having an Immortal time in physical time. Your immortality has to be funneled through a narrow tube called human experience. Imagine now funneling down from the dimensions of where you have come into a tiny compressed chamber called the earth and human experience. It's like being stuck in the birth canal. Your immortal self can't move; it is contracted and stuck.

You must give your soul space to breathe; it's your immortality into your human form. You must always remember that you — the real you — are having an experience in a human dimension (a third-dimension reality) that is terribly constrictive. Just remember who you really are and who you need to embrace in times of alienation and fear. The fear of being human is the fear of being stuck in this contracted time forever, and this is where humans panic, reach for stimulants, and kill themselves. They kill themselves in many ways when they don't acknowledge their immortal selves. Give yourself time in your day through concentrated breath and awareness to acknowledge that your true self is your immortal self.

Your feelings of limitation or being stuck now create in you panic in your inability and vulnerability to create all you want for your journey of immortality right now. You need to feel that you are really able to connect with your true self at this point. You need to find the flow and the freedom to really find that panic space and go beyond it to really merge in oneness with the Immortals. You need the freedom and expansion to do this and to really find the flow within yourself when you are panicky or isolated. You must seek to find that little space within yourself that really goes beyond your fear and alienation. Trust in that space now; really allow yourself to trust in yourself to believe in your truth. This is a time for simply allowing the serenity and trust to emerge, so you can find within yourself your own divine truth. Just feel the trust, and the feeling of panic will disappear.

Just say to yourself, *"I trust the flow of life; I know I am now stuck, and I seek to address this within myself"*. My ability to really go beyond my fear now allows me peace and fulfillment on my journey. I am in a space of peace and oneness now, and I do not fear this space of panic. In fact, I embrace it because it is my own fear of my immortality that I am embracing.

"I am embracing my immortality by really living in the present right now, and I am seeking to go beyond it by creating a space of oneness with all that is around me."

The trick is to simply feel the potential and passion of where you are at right now and seek to capitalize on the moment. The sense of passion and oneness is now all around me, and I create in this passionate moment. The passion is in the moment. Being passionate in the moment is the essence of true immortality. True immortality is living passionately in the moment and drinking from that moment, really taking it in. You need to feel passionate about the moment, really passionate, because it is passion that holds the key to life as it engages all the senses.

Engage the Senses

When all the senses are engaged, you then become attuned to all around you, and when you are attuned, you can create the magic alchemy of change. Anything can happen in this space. Anything can happen in this moment because all the senses are involved. Your senses become attuned to this vibration and, when trained, respond very quickly. Allow this passionate life of yours to envelope your every second, and you will be truly clairvoyant, truly magical, truly one with everything. The magic of alchemy — that is, the changing of one thing into another — now takes place. Engaging the senses in the moment creates the alchemy of change. The breath — conscious breathing — gets you there.

Engaging the senses is the most exciting opportunity that comes with being an Immortal Woman. To engage the senses requires the sensitivity and sensuousness to experience *all* around you as a transformative experience. You are transformed when you engage in the senses in all that you do. Engaging the senses is the joy of living in the moment. To engage the senses requires practice, and to do this, you are allowing yourself the opportunity to slow down your inquiring mind — your intellectual mind. Attune to your intellectual mind. Really talk to it as a friend. Co-create with it. Ask it for assistance in your new discovery of yourself. To do this, you need to find a moment to reflect. Say to yourself, "I am giving myself 10 minutes to really engage the senses to develop my ability to see myself as an Immortal Woman." Just sit and engage the senses. Really engage them and seek to find a space for this engagement.

Evoke the senses and really connect with the breath in co-creation with the senses. Now just experience that moment of surrender to the senses; evoke the sense of smell; let yourself drift away on the waft of scent—incense oils, the more subtle the better—and travel with this one smell. Really feel the texture of the smell. What is the smell communicating to you? Try and evoke as many sensations in merely connecting the scent. Try and get inside the smell and create yourself in the scent.

Really let yourself experience the scent and become one with the sense. Engage your olfactory senses now. You can engage all the senses in this way to really allow yourself to embrace your immortality. You may like to try to really allow yourself to be enclosed by the senses. The senses are a trigger—a metaphor for immortality—because they are creating in you the need to really allow yourself inner space to explore your true self. The true self is this exploration. The true self is the essence of the senses; the senses are the essence; the senses create the essence. The senses are the key to immortality because they engage you and take you out of yourself and really evoke in you the need for total release from the structure of being human. The structure of being human is in the release of the very structure you have created by being human in the first place. The subconscious mind is the first key to immortality because it is the very part of you that is keeping "you" a prisoner. You are being kept a prisoner of your subconscious mind, and you need to find the key, the missing link. Doing this is fun. Engage the subconscious mind now and tell it you are going on a journey with it so could you please have the key that will allow you to really explore the other part of you—your immortal or soul self—and tell it you want freedom to roam and play with your immortal self.

You need to roam and play with your immortal self now, so give it a go. Just engage the immortal soul to create with you the spark of truth and excitement for your journey to immortality.

To be immortal is to be free.

Create moments to deliberately engage the immortal self by playing with the immortal self. You can engage the Immortals themselves in their essence to come and assist you in your playful journey to discover this true/real part of yourself. This true part of yourself grows under the guidance of the Immortals themselves. The Immortals have a belief and knowledge of exactly what is required for you in human form to ascend and create with them. You need to find the pattern of the breathing sequence that brings them in. Try this now. Say, "I ask for assistance from the Immortals to come to me now and help me create true immortality of self." Breathe in slowly, deliberately, up to the count of eight, slowly deliberately, and lovingly and always evoking the light of spirit, begin slowly now. This is the frequency that gets the Immortals.

The frequency that keeps the Immortals will be maintained by this deliberate breathing sequence. Do this now; breathe in, hold, breathe out, and allow the sacred light of the true colors of The Immortals to ride on the sacred breath. Breathe in color with your breath now.

I am a winged, rainbow scarab flying to meet the Immortals now. Fly like the winged scarab now. I am now a winged flying scarab. I am a rainbow one; the wings are rainbow.

To create an Immortal experience while being human is to constantly challenge and monitor the human self—to go beyond the "normal" definitions of what being human is. This is to see yourself, witness yourself constantly in your emotional, energetic relationship with yourself; to do this is to create a "now" space in all emotional reactions to life's events. To create The Immortal One, the true essence of immortality, is to be a living, embracing energy of spirit over will and spirit over emotions. The spiritual or soul self will be now able to create all there is. This is all there is.

Emotional Enslavement

Witnessing your emotional reactions to life's events requires discipline and self-awareness. This is an important discovery about the self. The self wants to have the experience of growth. It does want to grow up. A child does not want to stay a child. It sees the adult world and dreams of being part of it. It wants to grow up. Your emotions keep you as a child. Your childish self is always emotional, and this is dangerous as it is keeping the soul enslaved in emotions. It cannot evolve in this state, and it cannot accept itself. It just cannot merge into the totality of its true sacred self. The emotions are dangerous enemies of the soul self, and this danger created in you is the danger of enslavement; to be enslaved by your emotions and not seek to go beyond them is to be a victim. And not to challenge yourself daily to release emotional extremes is to be enslaved by mortality. This is what mortality is: enslavement to emotions.

Creating Immortality

What exactly is creating immortality? To become a creator of immortality is to really connect with the divine within you and to really allow yourself the chance to bring to your life your belief in your truth and oneness in all of life.

Creating the message of oneness conveys a sacred truth and light for all of life and all of creation; your life and the essence of the Immortals give you all you need for your journey of being human. Just feeling the magic and essence of the divine Immortal Ones is the challenge for you to embrace; creating this challenge brings to your life your divine truth for your life of trust. Caring for the journey of being human is very important in the quest for immortality.

Just recognizing your divine truth and the Immortal Ones creates a sense of wholeness and love; the essence of wholeness and love just creates the magical spark of creation, which makes all of life a celebration of uniqueness and love. Creation and love bring the life of the Immortal Ones to you; allow the essence of creation and love, the vibration of truth and power. Bringing truth and power to you now is the essence of the Immortal Ones; allow this to happen now.

Allowing the sacred trust of the Immortals to be infused into your cellular makeup is a matter of simply acknowledging to yourself that you want to be an Immortal. You need to simply reprogram your cellular makeup to accept this new aspect of yourself, and your journey in human form will become a journey of self-discovery. To begin the journey of self-discovery with the Immortals now challenges you to make changes in your life; you would ideally need to program yourself every night before sleeping to give you the Immortal frequency and coding. You simply ask your cellular memory to accept new frequencies.

I now command my cellular memory to accept new frequencies.
I am asking for immortality to become part of my everyday world.
I am now an Immortal.

Record your experiences in a diary, and don't be afraid to challenge all core beliefs you have about yourself to be brought to your awareness. When they are brought to your awareness, you can simply transmute them and command that they *not* be part of your everyday experience.

Your life now holds the key to your immortality. Your life is a creation with the Immortals now, and you are being attuned to the vibration of self-trust and self-love; you are an Immortal. You are co-creating with the Immortals when you don't see your life in a linear way. You will now see your life in a universal way, and everything you do for yourself has consequences, which go beyond time (your time) and place.

The Immortals are very aware of the quest in human form for them to become part of you and you them. You are able to really find this frequency when you deliver to yourself all aspects of yourself to create in oneness with the Immortals because connecting to the Immortals is just that: creating oneness and all in one. We can be immortal when we really acknowledge to ourselves that we are immortal. I am an Immortal Woman, and why not. The status of immortality is a creation, and oneness with all there is; you need to feel that the Immortals will bring the divinity of you to yourself. To bring the divinity of yourself to yourself, you are now challenging all previously held beliefs about your core identity, and you need to feel the excitement and energy of your core identity melding with the Immortals. To believe and trust in the Immortals creates in you a belief in yourself that transcends time and space, and you will realize that your identity will be reshaped entirely. It's important that you do this; reshape your identity and allow yourself all you need for your life of self-trust. To be able to trust the self to be an Immortal is just that: trusting in the self to give you all you need for your immortal self. There is the totality of you having an experience in a human body at one given time. This gift to yourself is a most precious gift indeed.

To create in love is to create a sense of peace, oneness, and true belief in your immortal soul self: to really allow the essence of your true self to become one with your totality. Now is a time for the truth of your soul to be made manifest to you. This is a time for remembering who you really are and forgetting who you are in your human form. It is a time for just forgetting who you are in your inability to really connect in love to your true source of love and light. Allowing the oneness of spirit to be infused in your very being makes you really connect to the immortal aspect of yourself. This is the immortal soul self. The immortal soul self makes the connection in oneness with all of light, and all of light blesses the humanness of the soul that is struggling for its identity in a human body. This is a time for just connecting to the truth and allowing the truth of your immortal soul to be with you at all times. This is a time for completeness in spirit and oneness with all there is and a sense of oneness and peace with all there is. This is a time for completeness and oneness for all. Completeness and oneness in the immortal soul self brings The Goddess of All Light through the Immortals to you. You are in complete oneness and harmony with all in the adventure of the Immortal Ones. To create oneness, peace, and order out of sadness and chaos is the task of the immortal/soul self. Allowing yourself this task is the demand of the soul in the human body. The human body and soul are really one; the human body and soul create the one. The immortal soul creates this. This is the task of the immortal soul.

"Always an Immortal; always an Immortal." Say this to yourself daily. "I am always an Immortal; I am always an Immortal, and I truly begin the quest for my immortality right now – right in this time and space. Right now in my life is where I begin." Beginning this quest for truth starts with the self; the immortal self will begin to grow like a seed inside you. It needs feeding every day with light, love, and peace. The immortal soul speaks of this truth – this oneness – now. This truth, this oneness, this complete merging into oneness now, all is safe in my immortal world. All is one in my immortal world. I am an Immortal now embracing my truth.

Just remember that you are creating your own immortality every time you redress the past. If there is a past that is hurting you, you must seek to find the answer within yourself and seek to forgive that part of yourself now. You need to find fault with no one – only yourself. I am finding fault with me; I am allowing that part of myself that is disconnected to become connected again. You need to find the true forgiveness of self through just letting the past dissolve and merge into *nothingness*. To let this past slip away like a boat, just drift away, is the most important task of being an Immortal; you need to feel a sense of compassion for yourself in doing this, and this is the hard part.

Acknowledging Immortality

This is a time for acknowledgment of your true purpose in human form and acknowledgment of your gift to yourself in seeking a human incarnation. This is a time to just feed and connect to your own unique sense of self-worth and love and, above all, peace for yourself as a human, acknowledging your immortality.

Acknowledging your immortality is also acknowledging your mortality as well, because you are mortal first and foremost. You are mortal first. I am a human woman. Stop. Remember your humanness and remember to get your human needs met. Always meet the demands of being human. Always meet the demands of being a human first. Ask yourself, "What do I need to be human?" You need to feel strong and have a respect for yourself as a human. You need to feel freedom as a human, and, above all, you need to feel that you can create in your life your love of yourself. Just always connect to your humanness first—always put *me* first; always focus on your ability to really feel good about yourself every day.

Self Abuse

People become emotionally unstable and have mental health problems because they fail to acknowledge their worth. They are always feeling unworthy. They fail to get their core needs met first; they really fail themselves and fall into the degrading trap of self-abuse. Self-abuse is the byproduct of an uncaring person and an uncaring person cannot care for others or her environment. So self-abuse is the problem behind all mental and emotional disorders. The mind just gets fragmented and breaks off; it becomes disconnected and dysfunction results.

The immortal soul and the immortal body are one, so they should be *compatible*. There needs to be compatibility in everything. See the immortal soul and body as having to live together in a household, like two people sharing a life together. They have the same common purpose: to find, within the difference, harmony; there must be a dance of reciprocity between them, and there must also be an acknowledgment of the differences and a respect of what the differences mean to each aspect of the person in her soul self. The body is the slave to the emotions and instinctual demands placed on it by its human form. A human is a human, just as a dog is a dog. There are aspects of humanness that cannot be ignored. The soul, in its purity and timelessness and through forgiveness has to marry its evolution in the framework of the human incarnation.

They have to both mutually respect each other. There has to be an acknowledgment of that respect just like there has to be between two people living together—together but apart, together but different. You need to recognize that the soul needs time to be listened to, to be given space to grow in a human body, but the body must always feel the need to grow as well. We grow old to return to our soul self, so we must keep monitoring the body and its desires, impulses, or fears and not give in to them because the soul is enslaved to the body and will not grow if the body doesn't grow. So allow yourself growth in body, mind, and soul. It is then that you will begin the journey of immortality because you won't become fragmented or lost.

Chapter 10

BLISS

Oneness

Always in your life you are realizing that the immortal self requires love and respect. The love and respect an Immortal soul wants creates in you the need to really allow yourself your truth to emerge in oneness with all there is and to feel this oneness in everything. The oneness of everything is a belief that you are never alone and that aloneness is a space of denial of the true potential of the soul self. To create this separation and loss is indeed a tragedy, and this is why your soul is a fragmented one; this is why you are now enslaved by the body's demand to really take control.

Framework for Self Love

This is a time for acknowledging that you are now able to connect and love that aspect of yourself that is fragmented and lost. Acknowledging this creates a framework for living and a framework for being human. A framework for being human is just that: a framework. All humans need a framework for being human. There are certain things a human needs to be human, and that is to really know the recipe for self-love. What is the recipe for self-love? What are the ingredients for loving the self? Self-love requires different ingredients measured out in different proportions. Too much spice kills the taste; no spice creates blandness and no excitement in life. The recipe for living as an Immortal goes like this:

Absolute trust in the oneness of everything and that you are "never" alone. This is called the *cake*. This is the universe: the cake. Everything is one; nothing is separate; everything needs another thing at certain intervals and certain wavelengths. The cake is the universe. This is the most important thing of all and must be reinforced.

The cake is the universe, and we are part of the cake, and we need to recognize that every action, every deed, every thought creates the cake. Your cake must be of your own creation. Imagine a wedding cake being not perfect for the creative day of a couple making a commitment to share and grow a life. The ingredients must be of the highest quality; they must be carefully selected and perfectly balanced, and the batter must be carefully cooked. Then you have the perfect cake. You need to recognize this parallel in all of life, starting with your body. The body must be kept in a perfect state; nothing must be put into it to create imbalance. It needs rest, recreation, and work, and it needs nourishment and beauty. When we treat our bodies as our world, we create a state of perfect balance within us so we can create that perfect balance without; this is a time for remembering this.

The Immortal Soul

The immortality of your soul speaks to you now. My immortal soul wants freedom; my immortal soul wants remembrance of what *I* am; my immortal soul/spirit now finds in you/me peace to journey into truth. This is the time to remember that all that has happened is a result of the shared, sacred truth of all of creation. This is the time to feel the love of the universe in your immortal soul. The immortality of your soul is in the love you give back to yourself. The love you give back to yourself speaks to you from your soul; allowing the soul's essence to be one with you creates the spirit of immortality in everything. This is the allowance of spirit to merge into matter, and your journey is one of sacred trust in this sharing. The sacred trust in sharing your "self" with your immortal soul is the most important thing you can do on your journey. The importance of your journey can only be measured by how much your soul self has grown. This is a time for this remembrance and acknowledgment. I remember and acknowledge my soul's journey. My soul's journey is my journey. I must acknowledge to myself my soul's presence and seek to speak to my soul daily, hourly, if possible. This is a self-realized life, one of trust and acknowledgment of my journey into spirit. I am a spiritual essence living off the human. The human is only the vehicle for the spiritual essence of me, and I am now feeling the pull of all of creation in this truly. To experience this essence is to be one with all of life; it is the true experience of being human.

Believing in the immortality of yourself must first be an acknowledgment of your soul's purpose. Be practical about what you can do daily for your soul's purpose. My soul's purpose is to nurture the *sacred essence* of who I truly am. Who I truly am is the essence of all life in its totality. There is absolutely no fear of being human when you embrace immortality, just the knowledge that you are the soul made manifest in human form and that this human form has a duty. The duty is to the soul; that is my soul's duty now. The soul's duty is to really connect to all of life and embrace that life now. My soul's duty is to embrace the absolute essence of who I am and live that essence. This is a profound time.

Remember who you are and what you can do for yourself now; just remembering and connecting to all of life through the sacred breath makes you immortal now.

Creating a space to be an Immortal creates in you immortality and confers immortality upon you; you are truly one with the Immortals when you do this, so you must allow yourself time to grow this aspect of yourself now. You are an Immortal having an experience being human, so every day, try to understand what being human means to you today. Ask yourself, "What does being human mean to me at this very moment? What makes me human?" Then tell yourself, "It is my ability to learn to love that makes me human now." You know that you have the ability to really seek love for yourself by finding it in others, always. Others who challenge you mirror back to you an aspect of yourself that isn't human because "being human" is to be mirroring yourself to yourself. By mirroring myself to myself, I am being human. I am in the space of experiencing humanness, and I need to feel the energy, power, and life force of my humanness. Truly, to be human is to allow yourself this experience of self-love and self-forgiveness; you need to focus on just allowing this aspect of yourself to be one with you now. I am now one with me, and I am able to mirror my fears, anxieties, and frustrations to myself; I know that to be immortal challenges me daily to see humanness as a merely transient space, and I challenge myself to do this now.

To be able to really see the immortal part of yourself is, first, to recognize that you as a human soul will be present and watching everything the body and mind and emotions go through daily, and this witnessing can be torturing for the soul. In witnessing this truth, is the knowing that *you*, the immortal soul, is observing a part of yourself in a space of self-torture through the self-abuse of being human. Human self-abuse is the most devastating thing for the soul part of the human to witness. The witnessing of this destroys a sense of trust that the soul experiences. The soul experiences daily torture when you decide to self-abuse. The soul is like a child in many ways. It cannot express itself in a space of separation. Separation and trust don't exactly connect when you are feeling the weight of being human. Ask yourself whether you would abuse your own baby. Your own baby self is being abused when you are cruel to yourself because it separates you from your true self: the soul self. The soul self needs support to grow and nurture its identity in a human body; this is essential for its growth and development. It's just that simple. Souls need nourishment in human bodies, so nourish the humanness of yourself and nourish the aspect of yourself that needs to be reminded of this. I am nourishing the aspect of myself that needs nourishment by my soul self. My soul self is now being lovingly nourished by my attention to detail in all of life. All of my life is an acknowledgment of this aspect of myself that needs to be reminded of my immortality.

You now need to allow your immortal soul to really be present in all you do and say, and you need to be aware of this invisible presence and acknowledge this part of yourself daily. It is important to really allow this to happen and to really create all the conditions for this happening to become part of your everyday reality. Your truth is your beacon, and your truth now is to surrender fully to that part of yourself that needs this nurturing of spirit. The spirituality is just that: it is the nurturing of that part of yourself that is spiritual. True spirituality is the sensation of allowing the spiritual essence of yourself to truly merge in oneness with you; that is spirituality. Spirituality is the acknowledgment that your immortal soul is part of you too. It's like a responsibility. It is being responsible to a part of you that you need to feel is part of your essence; you need now to really acknowledge that part of yourself and let yourself flow and merge with it in unity for all of life. You are one with your immortal self in everything. You need to focus on this mergence when you feel the need to bolster yourself — when you are down. You need to really consider your immortal/soul self and ask it what it wants because you and it are one and the same. It's like asking your partner what to do about a decision that is life-changing.

Soul Breathing

Breathing always confers immortality because the soul can speak to you through the breath. The soul is carried on the breath, and the breath is the messenger of the soul. True *soul breathing* must be part of your everyday reality. Soul breathing is the acknowledgment that your immortal soul has a right to speak to you through the breath. So allow the breath to be the vehicle for the soul. Soul speak is sacred; soul speak will train your subconscious so that you are ready for soul speak. A true merging takes place when you *"soul speak."* Breath work is soul speak, and you need to train yourself to do this daily. Give yourself 10 minutes daily in soul time. It's not too much to ask when your immortal self wants attention; and what does it want attention for? It wants attention to live in you. You need to feel the power of simple purity and love for your immortal soul self. This is an exciting time just to be able to really connect in love with your soul self. This is the energy of pure trust and love for all of life. Looking at all of life is able to be convened on you if your absolute truth can be acknowledged through soul speak. Soul breathing is soul speak, which equals immortality. So just believe in the truth of your life right now; this is a time for honoring the forces of all the worlds to create with you.

Human Essence

The true spirit of immortality is the essence of being human because you are not really human without it. This is the truth of your journey right now, and this journey will give you all you need to really attract the right type of people, events, and situations to you right now. You need now to feel the strange energy of immortality around you because you have submitted to the will of your soul. This is your true truth; and your soul is your immortal self. Feel the deity's nature and all that you trust your soul to be part of now growing your immortality. This is a time now for the immortal soul to really connect with you in love for all of life; so bring to your life your essence and radiance for all of life in its immortal form.

Truth, peace, and surrender bring to you your immortal soul. Immortality is recognizing that you and your soul are one and that you and your soul will speak the same language: the language of immortality. True love and peace for the immortal soul is now bringing to you all you need for your life of truth. This is a time to just release and speak to the immortal soul through everything. Everything is one, and the truth of who you are now is the joy, peace, and magnificence of all you are capable of. I am now in a space of light for my immortal soul to grow.

Allowing yourself space to grow your immortal soul is your biggest challenge because you are struggling with the enormity of immortality, and this struggle is in itself the essence of immortality. The essence of immortality is the surrender to the soul and a surrender to your magical truth of exactly what you have created. This is a time now to remember all you have chosen and will choose in the future; your immortal soul is your only reality because everything else is an illusion and a mere mirror of fear, disgust, disaster, and abandonment. All of these energies are in competition for your immortal soul and seek to claim the soul for their development. You need now to really consider what you had and you need to feel the presence of the Immortal Ones now. The presence of the Immortal Ones can be called upon for assistance immediately; there is one of the above-mentioned energies; the soul is on a perilous path in human form, and you need to recognize this and seek to go beyond this.

The time is now to let yourself surrender to your true immortal self and fly your wings, knowing you are protected. All of life is about movement; nothing can stay still or it stagnates, becomes polluted, and rots; you just rot away if you become polluted.

Being in the presence of the immortal ones is an absolute joy, for you can release all to them. You need to feel that you can release to them and that they—the Immortals—are really allowing you to connect to them in divine love and peace. This is the time for allowing this aspect of yourself to be developed. You are radiating this acceptance of yourself when you believe in an aspect of yourself that needs developing.

Aspects of the Self

Ask yourself, "What aspect of myself do I need to develop today? What aspect of myself needs this growth?" It is very important that you connect to all aspects of yourself and grow this aspect of yourself now. You need to connect to this aspect of yourself and let yourself experience true joy; you need to feel the wonder and essence of all around you and feel completeness and peace in your quest for immortality. You need to really consider all you can when you connect to your divine truth through the Immortals. You need to feel the experience and the joy of your immortal soul talking to you now. This is a time for telling yourself that you need to consider the essence of your immortality, which lies inside you. Believe in the essence of your immortality right now and know the very best is open to you through your belief in your immortal life.

Divinity

Divinity through the immortal process is one of great power over your life now. Just considering your own divinity and power over your life creates immortality in all that you do. Now is the time for your essential self to really embrace this divinity and remembrance; your essential self has absolute joy and surrender in all of life. Simply surrendering to all of life is the essence of being human, and this essence of being human means that you allow the sacred presence of your immortal self to be one with you right now. Now is the time for this mergence of oneness and truth, and you need to feel this mergence take place. The essence of The Goddess of All Light creates the seed for immortality to grow, and you need now to really allow the seed of a divinity bigger than yourself to grow in you if you choose to become an Immortal. This is a time for allowing this divinity to take place within you and to nourish the seed of divinity and power within you; just allowing the special seed to grow a new you in the trust of yourself now is the creation of your immortal self. The self seed of creation must come before everything. This is a time to flow the energy through you for your immortal self, and this immortal self does not wither and die. Just allow the seed of immortality to be part of your identity now and claim this inheritance as your own. Just do this now, and immortality is yours.

Remembrance

The seed of immortality starts with the process — the process of remembrance — and the process of remembrance begins with the knowledge that you — your immortal self — now create the journey on which you will create the new and special in you. This is a time for owning your life and claiming your heritage in your immortal/mortal self; the essence of mortality is to live now — to create now and to be now. It is a fact that this knowledge will sow the seeds of growth for the immortal self. You are allowing the immortal self to be part of the journey of self-love. The true essence of loving the self is in the knowledge that you, your immortal self, will need reassurance of your commitment to the journey to claim immortality. This is a time to claim the totality of who you are and to feel the reward of this heritage. The totality of this heritage lies in the understanding of yourself as a totally new being, not one of the usual fear, pride, and anger you are always carrying. The fear, anger, and pride you are carrying now creates in you all that you need for you life of bliss and wholeness. The journey of creation begins to really claim you now. Your journey of creation is your own special link with your future, and you need now to feel the splendor and abundance truth brings. You need now to really claim your truth and live it. This is a time now for this remembrance; just remember *who* you are.

Immortal Spirit

Allowing your immortal spirit to merge into oneness with you now is a gift of life, and it is a precious one indeed. The immortal spirit speaks of freedom; the immortal spirit speaks of love and surrender; and it is this love and surrender that allows the essence of oneness to be made manifest. This essence of oneness is now a connection with your divinity, and the essence of oneness creates in you a specialness and lightness; this lightness and your love for yourself now is the essence of all that you are and all that you will be — the essence of yourself. It's this powerful manifestation of all you are; all you are is light, and it is this light that brings to you now your belief in your truth and power; all of love is in light and power and you now need to really nurture this holy spirit. Your spirit is holy, and it is made manifest in all that you do right now. It is a time now for remembering this space within you and fostering and generating all the love your soul can provide for your journey of truth. The journey begins with yourself because only you can walk on the road. Only you can bring to your life your holy spirit into your body, which must be whole.

A whole body brings a whole holy spirit, so first, the body must be attended to get it right, to have it functioning properly. Look after it like a thoroughbred animal that you want to win races; train, nourish, protect, nurture, and feel the body. Talk to it daily. Ask it, "Body, what do you want from me today?" Nourish the body; you can deal with the emotions.

Allowing the space and peace for the immortal ones to be a living part of you really challenges you now to connect in spirit and wholeness with all there is. This energy and life force brings your immortal spirit into resonance with all of life, and you now feel the magic energy and life force with all of life. This is a time to just let go of all fear around change and bring the energy, life force, and magnificence to you now. This is a time to just surrender all to yourself and feel the absolute joy and love in all of life. Life offers rich rewards for your journey. Now nourish it. You have come so far in your journey; it is important to feel this and create new heat and fire in your life right now.

Reconnect

All of the universe cheers you, all of the universe applauds your magnificent spirit, and all of the universe makes your life whole now. This is a time to remember all and to release to the essence of everything your whole life creates in you now: a desire to reconnect with your absolute, divine truth in everything and to live this wholly. This is a time to reconnect to all of life in its magnificence, wholeness, and joy; just be alive to the wonder of creation and bring this wonder to you; your whole life now shines, and you feel that the fantastic energy of being human is worth it.

The Dance of Passion

The love of yourself is the real self speaking. The passion for life is the real self speaking, so do this now: allow you life to gather passion and momentum. All of life is a dance, and you must embrace passion and live it. Your life is one of absolute joy and passion for all of life. All of life is a dance of passion. List your passions now; list what makes you feel passionate, and seek to further this passion in you at every level. You are able to embrace all aspects of your passion by recognizing that life is a story of passion. All of life embraces passion because, without passion, no life can begin. Look at how you were conceived — through energy and passion. The energy ignites the divinity of your spirit, and the passion is the fuel. Every act of creation is a passionate one and can be incorporated into all of life. All of life is a dance of the passionate aspects of yourself, and you need to feel the love and passion for all of life embracing you now. Dance the dance of passion, and you are dancing the dance of youth and freedom. There is no such thing as age when passion becomes engaged because it is energy.

All of life engages passion; seek to do this now, and allow your immortal soul/spirit to be really connected to your divine, immortal self. This is a time to allow this sense of passion to be with you and to really accelerate your life path. I develop my relationship with my passion.

The immortal spirit rides on every thought, and every thought is one of love and respect for the self. The self holds the key to immortality because the self creates the immortal space within and without. This is a time to really find the immortal soul in everything you do; the immortality is in the connection with all around you, and this is the truth of immortality. So ask yourself daily, "How can I connect with all around me now? How can all of my connections be meaningful to create in me immortality? How can my immortal soul be free to really connect with all of life? How can my immortal self create all I need for my life of truth and trust in myself to connect always?"

Nature's Elements

Engage yourself with others, nature, and the elements. Look at the elements first thing in the morning. What are they saying to you? What is the meaningful contact today with the elements? This is a time for acknowledgment of the elements in everyday experience. The elements are living energies, truly living energies. This is a fact, so the element must be connected to take yourself out of your small, fearful world. Use your five senses to engage the elements. See, hear, feel, taste, and touch the elements. These are the most tangible reminders of immortality, and they must be targeted for your emergence now. Really link with the elements for immortality and allow your immortality to really be part of you now.

The Essence of Life

Allowing the immortal ones to be part of your totality now brings to you all that you need for your life of trust in all of life. Just living is an Immortal experience, and the simple fact that you are human gives the Immortals a chance to really be part of your human experience. All of life is an Immortal experience when lived this way. You need connection with your immortality when you are totally engaged in the total process of being human because, to be human, is to be immortal. This is a time for connecting to the essence of your immortality. All the essence of you lies in immortality. The essence of your immortal soul is now in the experience of trust in the oneness of all; all of you is one in your truth and immortality. Bring to your life now your absolute trust in immortality. This is a time for all the absoluteness of everything to be with you; you need to feel the energy and absoluteness of everything; all is one, and one is the essence of your totality right now. The essence of all of life lies in the expression of yourself at this precise moment in time. The absolute essence of yourself awaits you in the application of love in everything. This is a most important statement as it implies total and absolute trust in everything. Every living thing is in the application of what, when, and who you are; your essence, magic, and life force awaits you now when you believe in your power to trust the absolute essence of everything. Your life is a testament to this; live it now in absolute trust and love for yourself; bring to your life the absolute essence of everything. All of life is a trust process. All of life is a journey of love, power, and respect for all of life. Life is a journey. Live it with the absolute essence of your divinity now, and you and your life will all be one.

The essence, the absolute essence of being human, is to really believe in your ability to think outside the parameters of the square your reality imposes on you. Your reality is shaped by your experiences, and your experiences are your reality. Now it's time to pretend: just pretend that you, the real you, wants to explore this reality. When you were a baby, your reality was your crib, then cot, then room, then house. You went to school. Your reality was shaped and shared by your tribal group. Your family shared values. But now, this is changing. Your reality now includes something more; your reality now includes the immortal world. It is at this point that wonderment must set in. You must be shown how to wonder; you must be shown how to really connect yourself outside your reality. So you need a reality check; check your reality daily. My reality is shaped by my previous experience. Who shaped this? Your parents, the society you live in, world governments, and other people and facets in your environment. Your reality is not yours; it's actually someone else's. So you need to examine everything about your life and ask yourself, "Who shaped this reality? Whose reality am I tapping into? Am I tapping into someone else's reality? This reality is mine; this reality is mine because someone else created it." So toss it out, build a new reality, and ask the Immortals to give you a new reality. Reality is fear of change; reality is stopping me from getting the best for my life.

The Goddess of All Light through the energy of the Immortals is always present when you tap into the frequency of the Immortals. The Immortals are those who have answered the call in their totality: that a moment-by moment experience is the absolute truth for a life to be lived completely. The essence of your immortal life is the shared, sacred truth in all of existence. This is being given to you now, and this truth is yours. Now you must seek to answer the essence of the Immortals in you by being aware of their energy and presence in your life. The energy and essence of the Immortals is in a shared truth; a shared truth must be yours now because you are an Immortal.

You are now an Immortal, and this is where this book must finish because you now have shaped your own reality to create immortality. The Immortal Woman has finished, and this chapter in your life is now finished. Remember where you were when you began this discourse. Now it is finished; now you are an Immortal because a person who, in her humanness, reads this and honestly endeavors to believe its truth and absorb its truth is now finished with the pain of being human. Of course to *be* human is to have pain; to be human is to trust that there is a blue print for living, which must be observed if you are to live a self-realized life.

This is the blue print for being human; and being human is to know how and when to read it; use it when you need to connect to a part of yourself that seems separate from the Immortals and do not be afraid to feel your humanness. Love your humanness because it is your greatest gift to yourself.

Through the Immortals

References

Chapter 1 References

1. Rollin McCraty, PhD; Annette Deyhle, PhD; Doc Childre, (2012) "The Global Coherence Initiative: Creating a Coherent Planetary Standing Wave," *Global Advances in Health and Medicine.* Volume 1, Number 1, March 2012.

2. Rollin McCraty, PhD; Doc Childre, (2010) "Coherence: Bridging Personal, Social and Global Health," *Altern Ther Health Med.* 2010;16(4):10-24.

3. Rollin McCraty, Raymond Trevor Bradley, Dana Tomasino, *The Resonant Heart, Shift: At the Frontiers of Consciousness.* December 2004 – February 2005

4. See 2.

5. Global Coherance Institute. About The Global Coherence Monitoring System http://www.glcoherence.org/monitoring-system/about-system.html (Accessed March 2012).

6. The Global Consciousness Project, Meaningful Correlations in Random Data http://noosphere.princeton.edu/ (Accessed March 2012).

7. Transcendental Meditation, Modern Science Documents the Maharishi-effect
http://maharishi-programmes.globalgoodnews.com/maharishi-effect/research.html (Accessed March 2012).

8. See 1

9. Armour, J.A., Ardell, J.L., eds. Neuro-cardiology. (New York: Oxford University Press, 2004).

10. Institute of Heart Math. (2012) Heart Articles. An Appreciative Heart is Good Medicine,
http://www.heartmath.org/free-services/articles-of-the-heart/appreciative-heart-is-good-medicine.html (Accessed January 2013).

11. Kent, KM., T.Cooper (1974) The Denervated Heart, *N England J Med* 291: 1017– 1021.

12. See 9.

13. Ziskind B, Halioua B, (2004) Med Sci (Paris). Concepts of the heart in Ancient Egypt Mar;20(3):367-73. Cardiologie, 20, avenue du Petit Lac, 95210 Saint-Gratien, France.

14. McCraty, R., Atkinson M., Tomasino D., Bradley T. (2009) The coherent heart: heart-brain inter-actions, psychophysiological coherence, and the emergence of system-wide order. *Integ Rev.* 2009;5(2):10-115.

Chapter 2 References

1. Clerico, Aldo, Recchia, Fabio A., Passino, Claudio, Emdin, Michele, (2006) "Cardiac endocrine function is an essential component of the homeostatic regulation network: physiological and clinical implications." *American Journal Physiology - Heart Circulatory Physiology 290:H17-H29; doi:10.1152/ajpheart.00684.2005*

2. Phillips, Dr Tony, (2012) "Hidden Portals in Earth's magnetic field." http://www.nasa.gov/mission_pages/ sunearth/news/mag-portals.html (Accessed September 2012).

3. Norman, Anthony W., (2008), "From vitamin D to hormone D: fundamentals of the vitamin D endocrine system essential for good health." *American Journal Clinical Nutrition. Vol. 88 No. 2 491S-499S*

4. University of Southern California Pressroom. (2012) Scientists twist light to send data. Available from: http://pressroom.usc.edu/scientists-twist-light-to-send-data/ (Accessed August 2012).

5. Haas, Professor Harald, (2011) "Wireless Data from every Light Bulb." Annual Technology Entertainment Design (TED) Global conference in 2011, Available from: http://www.ted.com/speakers/harald_haas.html (Accessed June 2012)

6. Australian National University, Research School of Astronomy & Astrophysics. Cosmic thread that binds us revealed. http://rsaa.anu.edu.au/research/highlights/cosmic-thread-binds-us-revealed (Accessed July 2012)

7. The Template, the foundation ceremony, original innocence. (2011) (DVD) Jiva and Juliet Carter.

Chapter 3 References

1. Cann, Rebecca L., Stoneking, Mark, & Wilson, Allan C., (1987) "Mitochondrial DNA and human evolution," *Nature* **325**, 31–36

2. Mitchell P, Moyle J (1967). "Chemiosmotic hypothesis of oxidative phosphorylation." *Nature* **213** (5072): 137–9.

3. Jabr, Ferris (2010), "Gleaning the Gleam: A Deep-Sea Webcam Sheds Light on Bioluminescent Ocean Life." Available from http://www.scientificamerican.com/article.cfm?id=edith-widder-bioluminescence(Accessed June 2012).

4. Lemieux, H., Hoppel, C.L., (2009), "Mitochondria in the human heart." *J Bioenergetics and Biomemembranes. Apr;41(2):99-106.* Available from: US National Library of Medicine National Institutes of Health http://www.ncbi.nlm.nih.gov/pubmed/19353253 (Accessed: June 2012).

5. Narby, Dr Jeremy. (1999) *The Cosmic serpent: DNA and the origins of knowledge.* Jeremy P. Tarcher. Putnam.

6. Perlmutter David and Villoldo David, (2011). Power up your brain: the Neuroscience of Enlightenment. Hay House Inc.

7. Gote, Srikank. 2011, "Bacteria our future storage device." *International Journal of Computer, Information Technology & Bioinformatics,* Volume 1, Issue 3.

8. Lane, Dr Nick, and Martin, Dr William, (2010), "The energetics of genome Complexity." *Nature,* 467, 929–934.

9. Lane, Nick, (2006). *Power, Sex, Suicide: Mitochondria and the Meaning of Life.* Edition. Oxford University Press, USA.

10. *Star Wars, The Phantom Menace,* (1999), Film. Directed by George Lucas, USA, Universal Pictures.

11. Midichlorian, n.d. Available from http://starwars. wikia.com/wiki/Midi-chlorian(Accessed June 2012).

12. Ocala Firefighter Lifts SUV, Frees Crash Victim's Arm, (2008). http://www.floridatoday.com/article/20080702/ BREAKINGNEWS/80702008/Ocala-firefighter-lifts-SUV- frees-crash-victim-s-arm?nclick_check=1 (Accessed March 2012).

13. Wise, Jeff (2009), "When Fear Makes Us Superhuman, Can an extreme response to fear give us strength we would not have under normal circumstances?" http:// www.scientificamerican.com/article.cfm?id=extreme-fear- superhuman (Accessed March 2012).

14. Josh Clark. (n.d.). *How can adrenaline help you lift a 3,500-pound car?.* [http://entertainment.howstuffworks. com/arts/circus-arts/adrenaline-strength.htmAccessed 27 March 12].

15. Ikai and Steinhaus, (1961) "Some Factors Modifying the Expression of Human Strength," *Journal of Applied Physiology.*

16. Koltai E, Hart N, Taylor AW, Goto S, Ngo JK, Davies KJ, Radak Z. (2012) Age-associated Declines in Mitochondrial Biogenesis and Protein Quality Control Factors are Minimized by Exercise Training. *Am J Physiol Regul Integr Comp Physiol.* 2012 May 9.

Chapter 4 References

1. Armour, J.A., Ardell, J.L., eds. (2004). *Neuro-cardiology*. New York; Oxford University Press.

2. Dolphin Communication Project, So high it Hertz. http://www.dolphincommunicationproject.org (Accessed; July 2013).

3. McCraty R, Atkinson M, Bradley RT. (2004), Electrophysiological Evidence of Intuition: The Surprising Role of the Heart, Journal of Alternative and Complementary Medicine; 10(1):133-143

4. Skinner, Stephen. (2006). *Sacred Geometry Deciphering the Code*, New York, Octopus Publishing Group Ltd.

5. Luminet, J.P. *et al* (2003) Dodecahedral space topology as an explanation for weak wide-angle temperature correlations in the cosmic microwave background *Nature 425, 593.*

6. Study Mode, The Anatomy of the Eye and the Physiology of the Vision. (2011) http://www.studymode.com/essays/The-Anatomy-Of-The-Eye-And-584449.html(Accessed: February 2012).

7. Herz, Rachel S. *et al.* (2004). "Neuroimaging evidence for the emotional potency of odour- evoked memory." *Neuropsychologia* 42 (2004): 371-378.

8. Proust, Marcel. (1982). *Remembrance of Things Past*. Vol. 1. Trans C.K. ScotT Moncrieff and Terence Kilmartin. New York: Vintage.

9. Jenny, Hans. (1967). Kymatik: Wellen und Schwingungen mit ihrer Struktur und Dynamik/Cymatics: The Structure and Dynamics of Waves and Vibrations, Basilius Press.

10. Goethe, Johann, Wolgang. (1829). *Conversation with Eckermann* in: *Familiar Short Sayings of Great Men,* 2012, 6th ed., comp. by Samuel Arthur Bent. Boston: Ticknor and Co., 1887. www.Bartleby.com/344 (Accessed, November 2012).

11. Cymascope: Sound made visible. http://www.cymascope.com/cyma_research/musicology.html (Accessed: June 2012).

Chapter 5 References

1. Haramein, Nassim. To Infinity and Beyond: Transcending our Limitations (cont.) Energy Density of the Vacuum. http://www.grahamhancock.com/forum/Haramein (Accessed, August 2012).

2. Haramein, Nassim. (2009). The Schwarzschild Proton. Proceedings of the 9th International Conference CASYS'09. University of Liege, Belgium.

3. Haramein, Nassim, (2012). Quantum Gravity and the Holographic Mass. *Physical Review & Research International,* ISSN: 2231-1815 ,Vol. 3, Issue 4 (October-December).

4. Young, Arthur, M., (1976). The Reflexive Universe: Evolution of Consciousness. Delacorte Press, New York.

5. See 2.

6. Bentov, itzhak, (1977) *Stalking the Wild Pendulum: On the Mechanics of Consciousness,* E. P. Dutton,, ISBN 978-0-525-47458-6; Inner Traditions - Bear and Company.

7. Bentov, I. (1982) *A Brief Tour of Higher Consciousness: A Cosmic Book on the Mechanics of Creation.* Inner Traditions - Bear and Company.

8. Rollin McCraty, Ph.D. The Energetic Heart: Bioelectromagnetic Interactions Within and Between People. E book. www.Heartmath.org

9. See 2.

10. Bousso, Raphael (2002). "The Holographic Principle." *Reviews of Modern Physics* **74** (3): 825–874.

11. Bohr, Niels (1988) *The Philosophical Writings of Niels Bohr*, Ox Bow Press, Woodbridge, Conn.

12. Richard Feynman (n.d.) The Royal Society Invigorate, bringing science to life. What is quantum physics all about. http://invigorate.royalsociety.org/ks5/the-best-things-come-in-small-packages/what-is-quantum-physics-all-about.aspx Accessed: June 2012.

13. Black Elk, (1863 -1950), Teach Peace Foundation, Peace Quotes. http://www.teachpeace.com/peacequotes. htm(Accessed June 2012).

14. See 3.

15. Perez, JC. (2010) Codon populations in single-stranded whole human genome DNA Are fractal and fine-tuned by the Golden Ratio 1.618. *Interdiscip* Sci. 2010 Dec;2(4):373.

16. Scott F Gilbert. (2000), Developmental Biology, 6th edition, Early mammalian development. Swarthmore College, Sunderland (MA): Sinauer Associates.

17. See 3.

18. See 3.

ABOUT THE AUTHOR

Carmel Glenane B.A. Dip Ed. Owner/ Director of Atlantis Rising Healing Center™ and Mystery School. Founder of the philosophy of The Divine Feminine in 2002 and Senju Kannon™ Reiki in 2008, teaches The Divine Feminine Mysteries through her Mystery School Ascension Training program. A powerful interactive and dynamic motivational speaker, channelled writer, esoteric teacher, and sought-after healer, Carmel is known for her transformative tours to sacred destinations such as Hawaii, North, Central and South America, Turkey, India, Bali, Japan, Egypt and the great central heart of Australia, Uluru. With more than 20 years in business in personal development, Carmel's intent is to allow people to receive through the Heart's Intelligence through the mother's wisdom.

Carmel is the Australian Ambassador for HappyCharity.org as Director of Happy Spirits. She is currently writing training programs for all of her books to offer her courses online to a worldwide audience.

Feminine energy teaching programs became a focus after 10 years of founding my business Atlantis Rising Healing Centre in 1992, which led me to spontaneously become a channel for "The Goddess of All Light," who guided me to establish "The Philosophy of The Divine Feminine" in 2001.

Daily my consciousness is aligned to "The Goddess of All Light," where I receive written transmissions for my personal guidance and teaching.

Facilitating and leading my first teaching tour to Egypt in 2002; I have since taught in Egypt every year, as well as North, Central, and South America, Turkey, Greece, Hawaii Islands, Indonesia, India, and Australia. Each tour, book, and training has helped me "Earth" my body of Light; purging Earth attachments, as my ability to "Earth" (plug in) develops, holding more energy and light in my heart, for The Divine Ones to manifest.

In 2014, I was invited to open a Crystal Tones® Crystal Singing Bowl Sound Temple, and now incorporate these sonic Masterpieces into all my teaching programs.

I am in service to "The mother" and aim to have as many people as possible embody the teachings of our "Mother" through my books, teaching, and healing programs.

I am currently writing online courses to support The Atlantis Rising Mystery School ascension training program and creating new Guided Meditation mp3's to support the Ascension program.

About Lisa Malcolm Contributor to Part I Awakening the Intelligent Heart

Lisa Malcolm B.App.Sc. Lisa has a Bachelor of Applied Science and has been designing and presenting education programs within the university system for the past 10 years. She is an environmental scientist who sees that everything already exists for a truly sustainable way of living and to reach it we must make the journey into our hearts. On a spiritual journey of discovery from a young age, Lisa is passionate about the re-merging of science and spirituality and blends her backgrounds in both into holistic knowledge, research, and teachings.

To Connect with Carmel Glenane:
www.carmelglenane.com
www.senjukannonreiki.com
www.atlantis-rising.com.au
Ph: (+61) 0755 367 399

Recorded Meditations
By Carmel Glenane
featuring Crystal Tones Crystal Singing Bowls

New Dawn Meditation
By Carmel Glenane
Feat. Crystal Tones® Singing Bowls

Set Yourself Free Meditation
By Carmel Glenane
Feat. Crystal Tones® Singing Bowls

Core Identity Meditation
By Carmel Glenane
Feat. Crystal Tones® Singing Bowls

Today I am Receiving Love
By Carmel Glenane
Feat. Crystal Tones® Singing Bowls

Trusting to Receive Love
By Carmel Glenane
Feat. Crystal Tones® Singing Bowls

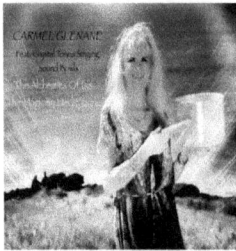

The Alchemies of Isis The Magician
Meditations
By Carmel Glenane
Feat. Crystal Tones® Singing Bowls

Embodying The High Priestess Meditations
By Carmel Glenane
Feat. Crystal Tones® Singing Bowls

The Immortals Meditations
By Carmel Glenane
Feat. Crystal Tones® Singing Bowls

Awakening The Intelligent Heart Meditations
By Carmel Glenane
Feat. Crystal Tones® Singing Bowls

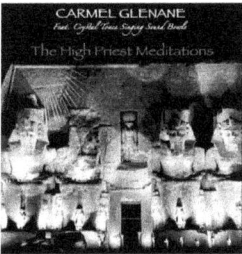

The High Priest Meditations
By Carmel Glenane
Feat. Crystal Tones® Singing Bowls

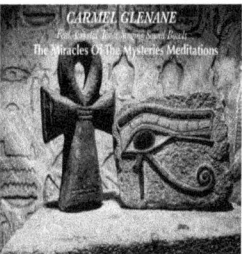

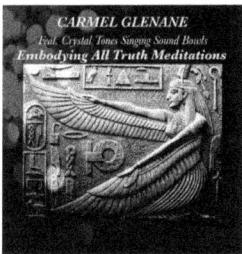

The Miracle of The Mysteries Meditations
By Carmel Glenane
Feat. Crystal Tones® Singing Bowls

Embodying All Truth Meditations
By Carmel Glenane
Feat. Crystal Tones® Singing Bowls

All meditations are available on these and other fine retailers:

iTunes

DEEZER

emusic

Other Books
By Carmel Glenane

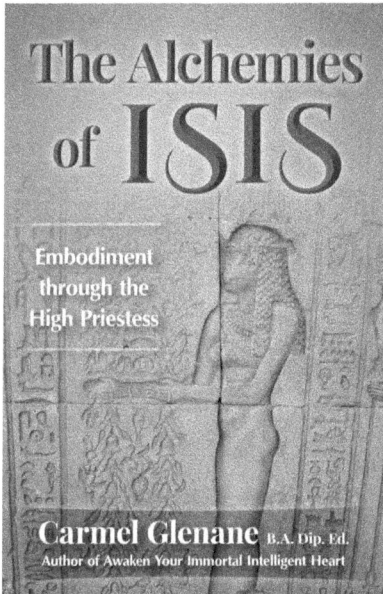

The Alchemies of Isis, Embodiment through the High Priestess By Carmel Glenane B.A. Dip. Ed.

Love fearlessly and passionately, for Love is timeless, infinite and unconditional!

Explore the Feminine energies with Isis and The High Priestess.

Carmel Glenane B.A.Dip.Ed. Author of several books on The Divine Feminine Mysteries has now combined *The Alchemies of Isis* with *The High Priestess* bringing readers the opportunity to embrace the secret wisdom of The Divine Feminine.

The *Isis* story is the story of all who love. Hope, restoration, and magic are yours when you lovingly embrace Isis in all her aspects. Every word brings Isis into your heart with "her" words of wisdom, power, truth and magic. Isis heals, restores, renews and resurrects new life; helping you open your heart to receive more love.

In the companion volume *The High Priestess* Carmel explores core issues in women's lives; relationships, intimacy, emotional love and spirituality in direct dialogue with *The High Priestess*.

You will receive Moon, Stellar (star), Nature Speak Meditations and Rituals for activating your 'Core Identity' to receive love, as well as lessons inviting you to deepen your relationship with your heart's truth.

"The Alchemies of Isis teaches us that every woman needs to be grounded and feel empowered, to be truly sexy and secure. I now feel both."

Dr Shelley Sykes
TV Host and award-winning author of Sexy Single and Ready to Mingle.

THE HIGH PRIEST

Author of
The Immortal Woman,
The Alchemies of Isis, and The High Priestess

CARMEL GLENANE
B.A. Dip. Ed.

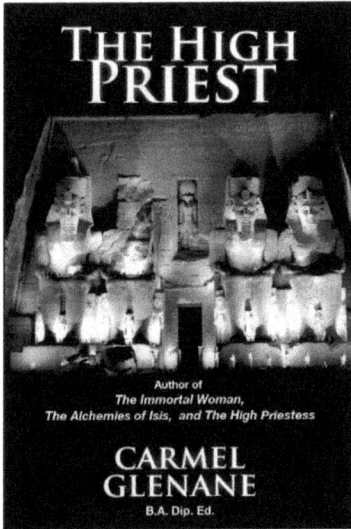

The High Priest, A Blue Print for Living through the Masculine Heart
By Carmel Glenane B.A. Dip.Ed.

ISBN: 978-0-9873439-2-5

FEEL WORTHY TO BECOME A MODERN SPIRITUAL WARRIOR, SEEKING TRUTH AND CREATING A POWERFUL NEW REALITY IN YOUR LIFE!

Carmel Glenane, internationally published author of 'Awaken Your Immortal Intelligent Heart' and 'The Alchemies of Isis' brings to you 'The High Priest'.

- Are you worthy to become a modern spiritual warrior?
- Do you want to be a truth seeker?
- Are you ready to activate your hearts intelligence?
- Are you ready to release your attachments?
- Are you willing to really manifest and rip out your illusions about your creative masculine?

In The High Priest, Carmel Glenane helps you explore and take control of your creative masculine. The High Priest offers a practical guide to living powerfully, with truth, changing your definition of your current reality. Join Carmel Glenane as she explores your creative masculine and discover your power.

Allow yourself your truth through power, to bring protection, power and love!

Carmel Glenane Training and Workshop Programs

Carmel offers training and workshop programs on the Activation of The Intelligent Heart in Australia and overseas.

Training programs are also offered for all levels of Senju Kannon Reiki™, through the Atlantis Rising Mystery School.

Carmel also facilitates tours to sacred destinations throughout the world. Carmel travels with Crystal Tones Singing Bowls and invites you to bring your own bowl to be a witness to your heart's remembering in all sacred sites.

If you wish to sponsor training programs please contact the Atlantis Rising Healing Centre™ office:
email info@atlantis-rising.com.au

For upcoming workshops or training locations or tour destinations please see our Website:
www.carmelglenane.com or www.atlantis-rising.com.au

⫷ *Transformational* ⫸
Tours

CARMEL GLENANE TOUR
FACILITATOR

EGYPT TOURS

The essence of Egypt is in aligning your consciousness to the Ancient Deities themselves. The Hieroglyphics of the temples reflect, through the reliefs, architecture and atmosphere the energies of the Goddess's & Gods.

Egypt feeds your soul. Imagine being initiated to the frequencies in the Kings, Queens, and Subterranean chambers also known as the Pit in the Pyramid of Giza. Reflect on the timeless wonder of the Sphinx, touching the stele at the heart of its initiation chamber between the paws.

Discover old Atlantis again at Sakkara, as the desert winds whisper their secrets. Horus the Falcon hovers as he guides you to his temple complex in Edfu, revealing the essence of order, protection, and freedom—the ancient Egyptians were known for creating order out of chaos. Float upon the Nile, which reflects the starry body of the Goddess Nuit. Re-code your cellular memory with The Great Father Osiris as he resurrects your weary spirit in this most ancient of temples complexes in Abydos, healing you by the green Osirian well, where the secrets of the Ancient Flower of Life can be revealed. Be drawn to the mighty temple complex of Abu Simbel where Ramses II and Nefertari's love was immortalized.

Allow our holy Mother Isis to enfold you in her wings of love as we sail to her sacred home on The Isle of Philae.

The magician High Priestesses and Priests of Karnak allow you to embrace your own magical powers in their home of balance and duality. Sekhmet the austere warrior goddess/mother of Karnak will receive you if you respect her power.

Explore where ancient rituals and offerings were given to the stellar forces at Dendara, home of the Hathors, Goddesses of Love and Pleasure.

JAPAN TOURS

Japan is a transformational, feminine and nurturing experience, especially Mt. Kurama known as the 'heart' of Japan. Mt Kurama, located 40mins outside the imperial ancient city of Kyoto, is the mountain where Dr. USUI received his enlightenment. Japan's delicate and very special spiritual energies reflect the beauty and power of Senju Kannon Mother of Japan and mother of our Feminine Reiki.

The Japan journey begins in the ancient shrine city of Kyoto, visiting powerful Buddhist temples, including Sanjusangen-do–The Thousand Armed Kannon Temple with 1,000 Kannon's (also known as Quan Yin) statues in the temple. The city's beauty is phenomena, featuring spectacular gardens, a geisha district, authentic Japanese cuisine, peace, order and tranquility. Unwind in traditional Onsen style bathing houses of warm mineral springs, relax in the peaceful retreat rooms or indulge in authentic traditional Japanese cuisine.

Our training program honors the Dr. Usui (traditional Reiki) but embraces 'The Mother' feminine heart of Reiki. Our programs are tailored to sacred sites and locations in and around the Temples at Kyoto and Mr Kurama. Travelling with our Crystal Tones Singing bowls magnifying and amplifying this incredible energy.